William Spotswood Green

Among the Selkirk Glaciers

being the account of a rough survey in the Rocky Mountain regions of British

Columbia

William Spotswood Green

Among the Selkirk Glaciers

being the account of a rough survey in the Rocky Mountain regions of British Columbia

ISBN/EAN: 9783337287023

Printed in Europe, USA, Canada, Australia, Japan

Cover: Foto ©Andreas Hilbeck / pixelio.de

More available books at **www.hansebooks.com**

AMONG THE SELKIRK GLACIERS

"The swelling slopes of forest, the blue ice of the glaciers, and the
dark purple precipices of Mount Bonney."

AMONG THE

SELKIRK GLACIERS

BEING

THE ACCOUNT OF A ROUGH SURVEY IN
THE ROCKY MOUNTAIN REGIONS
OF BRITISH COLUMBIA

BY

WILLIAM SPOTSWOOD GREEN, M.A., F.R.G.S., A.C.

AUTHOR OF "THE HIGH ALPS OF NEW ZEALAND"

𝕷onðon

MACMILLAN AND CO.

AND NEW YORK

1890

RICHARD CLAY AND SONS, LIMITED
LONDON AND BUNGAY.

I Dedicate this Book

TO

MY MOTHER.

PREFACE.

THE Paper I read last winter before the Royal
Geographical Society on our explorations in the Selkirks being necessarily limited as to detail, I thought
it might prove interesting to some to have a fuller
account. I have therefore in the following pages attempted, while describing our wanderings and scrambles, to give as complete a picture as possible of the
most striking phenomena of the region we visited. Our
map, now reproduced, was published in the *Proceedings
of the Royal Geographical Society* for March, 1889, and
it has been to me no small gratification to receive letters
from travellers in Canada, the United States, and at
home, men not previously known to me, saying that
they found it useful, and that it helped them to enjoy
the scenery of the Selkirks during the past summer.

Though all the time at our disposal was devoted
to rendering our map as accurate as possible, more
careful surveys will find many corrections necessary
in detail.

With regard to the comparatively small area surveyed, I can only plead limited time, and the difficult
nature of the country.   Without desiring to make
over-much of these difficulties, I think I may, in fair-
ness to ourselves, quote the words of one whose long
years of exploration in the Selkirk region, and whose
experience in other lands, give him a right to speak
with some authority.

Mr. Baillie-Groman, in his article in the *Field* of
May 11th, 1889, entitled, "Seven Years Path-finding
in the Selkirks of Kootenay," says, "The least said
about paths in these amazingly inaccessible upland
mountain wilds the better, for I doubt whether any
other known mountain system of the same not very
excessive altitude offers, on the one hand so many
attractions, and on the other hand, so many difficulties
to impede their exploration as do the Selkirks; but
that is to the genuine explorer only an additional
charm."   Again, commenting on the limited area we
were able to survey, "However, it must be said that
Mr. Green tackled that part of the Selkirks which
he explored from the very hardest and most un-
promising point of attack."   I regret that my meeting
with Mr. Baillie-Groman on the Columbia lake did
not take place before my work was finished; however
my object was more to throw light on the mountain
fastnesses in proximity to the railway than to attempt
a general survey of the range.

Of the illustrations, three, viz., the Frontispiece, the Snow Sheds, and the demolished Forest, are from photographs by my companion, the Rev. H. Swanzy. The others are from my own sketches.

In conclusion, I must take this opportunity to thank Sir W. C. Van Horne, and his Secretary, Mr. A. Piers, for their kind help when we were in Canada; Major Deville and Professor Macoun, of Ottawa, for their practical advice; Professor Bonney, since our return, for examining the rock specimens I brought home; the Council of the Royal Geographical Society for their material aid, and many others, too numerous to mention here, who "lent us a hand" when it was most needed.

W. S. G.

CARRIGALINE,
Christmas, 1889.

CONTENTS.

CONTENTS.

LIST OF ILLUSTRATIONS.

MAP.

AMONG THE SELKIRK GLACIERS.

CHAPTER I.

> " The food of hope
> Is meditated action ; robbed of this
> Her sole support, she languishes and dies.
> We perish also ; for we live by hope
> And by desire ; we see by the glad light
> And breathe the sweet air of futurity ;
> And so we live, or else we have no life."
>
> WORDSWORTH.

Introductory.—The idea suggested.—My companion.—Preparations.

WHEN the British Association met in Canada in 1884, one of the most interesting excursions planned for the members was that provided by the Canadian Pacific Railway Company on the portion of their line then completed ; to the summit of the Hector pass, or as it was then called, the Kicking Horse pass, in the Rocky Mountains. Amongst the members of that excursion were two gentlemen, Mr. Richard M. Barrington, and my cousin, the Rev. Henry Swanzy, who, not satisfied with the interesting scenes revealed to them

B

by the completed portion of the railway, determined
to continue the journey to the shores of the Pacific,
with the aid of pack-horses.

After separating from the excursion party on
Hector pass, they experienced very considerable difficulties. The temporary track for construction trains
was available only as far as the Ottertail bridge on the
western slope of the Rockies. From this point they
had to depend entirely on their horses. Having been
ferried across the Columbia river, they followed a most
imperfect trail, up the valley of Beaver Creek, into the
Selkirks and so reached Rogers pass. Often missing the
trail, they were compelled to make the best of their way
along the precipitous mountain side, through tangled
forest, until descending by the side of the Illecellewaet
river they rejoined the Columbia in the more westerly
portion of its course. They ferried once more across
its waters and on its further shore met the trail in the
Gold Mountains, which they followed to the shores of
the Shushwap lake. Here taking the steamer to Kamloops, they finally reached the railway at Spence's bridge
in the valley of the Thompson, and so completed their
journey to the Pacific.

The pedestrian portion of their journey was about
170 miles, and as there was always an uncertainty as
to what difficulties might lie ahead, they wasted no
time *en route*; but even so, owing to the imperfection
of the trail, it took them seventeen days from the time

they left the railway at Hector pass to reach the steamer on the Shushwap lake. Of the three pack-horses, two only survived the journey. The other poor beast, after numerous falls, became so disabled that it had to be abandoned seven days before the lake was reached.

Their return journey was made across the continent by the Northern Pacific Railway, and soon after their arrival at home, H., my cousin, related to me a full account of their adventures; his description of the great beauty of the mountain scenery of the Selkirk range awakened my interest and caused visionary desires to rise in my mind that some day or other, I too might have a chance of seeing those vast pine forests, with their grand background of glacier-clad peaks.

In 1886 the first through train of the Canadian Pacific Railway ran from the Atlantic to the Pacific, and since then, travellers have many times written accounts of that journey, the grandeur of the Selkirks invariably proving the climax of their wonder and admiration. I took for granted that the Selkirks were pretty well "done," now that a railway ran through the midst of them, and they faded by degrees from their place in my imagination. However, in the autumn of 1887 I chanced to meet a gentleman who had gone as special artist to the *Graphic* on the British Association excursion. He had lately met Professor Macoun, of the Canadian Geological Survey, and gathered that not only were the Selkirks as yet almost entirely unexplored, but

that it was much to be desired that some one who
had had experience of glacier-clad ranges should explore them. A very brief correspondence with Professor
Macoun settled the question, and I determined to go.
Then came the most serious question of all, in an
undertaking like that before me, Whom should I ask
to join me? Many a man whose acquaintance is most
valuable, whose help from a scientific point of view
would be great, and whose company on a short excursion
might prove delightful, would perhaps not be the one
to work with, when not only the body is worn out
with fatigue, but when the mind is too tired to pre-
serve its usual "civility" to outward circumstances.

Those who have had experience of prolonged expeditions, where men are inseparably thrown together,
and have to endure hunger and thirst, fatigue and
sleepless nights in company, will understand what this
means. Those who do not understand it had better
think twice on the subject before starting on such an
expedition. Men must know each other most in-
timately in order to combine with success, or else one
must be "boss" and the others subservient to orders.
H. and I knew one another sufficiently well to promise
a successful combination of this nature. We had travelled
together in Switzerland, and had attained to the great
stage of perfection—that of being able to squabble
with impunity. He had been in the Selkirks before,
and had recently been practising photography. I knew

him to be capable of enduring any amount of fatigue, so I was fortunate in inducing him to become my companion. We began by trying to find out what prospects of help we could count on in the way of porters to carry our goods in the mountains, and after some correspondence on the subject, I abandoned the idea of bringing men with me from Europe.

Some fine photographs taken from the railway track enabled me to form some idea of the nature of the mountains. Mountain camps should evidently be provided for, so I selected a small Alpine tent, made on Mr. Whymper's plan, which I had used in the New Zealand Alps; also a larger duck tent which had been our permanent camp on that expedition. I now got made a larger Alpine tent to hold four. This was formed on the same plan as the smaller one; the floor, sides, and one end being sewn permanently together. The material was unbleached calico, stitched by a friend in a sewing-machine, soaked with linseed oil and hung up for a month to dry. The test of two long expeditions have now proved to me that this is the lightest and most serviceable kind of tent. A curtain of mosquito net attached to the door makes it perfect. It is impervious to wind, insects and rain, and small objects are not in any danger of being lost, when laid carelessly on the tent floor. Felt sleeping bags, blankets, and a canteen, which had been with me in New Zealand made up our camp outfit.

The next part of our equipment to be thought of was fire-arms. What should we need? We heard of bears, bighorn, and grouse, but the chances of sport, when we were not going definitely for it, were not very great. Our armament finally consisted of a Henri Martini rifle, taking the ordinary army ammunition; a very light little Snider carbine, an Express, a twelve bore double-barrelled gun, and a light walking-stick gun which fired either shot or ball. This latter was thought to be handy for providing animals for the pot, but we afterwards found out, that it was rather more dangerous to the person who fired it than to anything else. We did not imagine there would be need for such a startling armament as this, but not knowing which weapon would be most useful we took all.

Coming nearer to the special work of our undertaking, we had to plan our photographic arrangements. My cousin undertook chief charge of this department, and he brought two cameras, one for quarter-plates and one for half-plates, and I packed up another half-plate instrument and a small Stirn's detective camera, which latter was a most useful adjunct, and did most satisfactory work.

Sketch-books and water colours were added, and then came some apparatus lent by the Royal Geographical Society, which may be considered as "the chief cause of our existence"—the surveying instruments.

These consisted of a plane table, legs, and aledade, which packed into a knapsack case and weighed— as our shoulders knew to their cost—no less than 27 lbs. I had a lighter table made to fix on my camera stand which we often used instead; and sometimes we fixed the camera by a special screw on to the head of an ice-axe stuck firmly in the snow. I tried to attach the plane table in a similar manner, but the difficulty of levelling and steadying were too great to commend the plan. A prismatic compass, two aneroids, a set of thermometers and a hypsometer made up the list from the Royal Geographical Society.

Besides the above I took a six-inch sextant, which unfortunately came to an untimely end. Fifty sheets of paper cut to fit the plane table were packed in a tin case and an ample reserve stowed among my luggage. I sometimes used both sides of these sheets in this way. Having marked off the angles on one side I turned up the sheet, sketched the panorama, and numbered the peaks in view with figures corresponding to the bearings taken on the other side.

As the first thing necessary in anything of a trigonometrical survey is to be sure of a base line, and though I hoped that such a line might be found already provided by the line of railway, I brought a steel line 220 yards long ($\frac{1}{8}$ of a mile) accurately measured before leaving **home**; **this** saved us much time in measuring.

Thanks **to helping** hands of kind friends, all the gear was ready **and** everything settled **for** us to leave Queenstown, for **New** York in the *City of Rome* on June 29th, 1888.

CHAPTER II.

" All things move Westward Ho . . . . It is bound up in the heart of man, that longing for the West."—KINGSLEY.

The *City of Rome*.—New York.—The Hudson.—Lake George.—Ottawa.

ON Wednesday, June 28th, H. met me in Queenstown. We gave our goods to the agents of the Anchor Line Steamship Company, spent a last evening with our friends and next morning went on board the tender, which conveyed us outside the harbour where our big ship was at anchor waiting for her passengers.

For two days after leaving Queenstown we enjoyed splendid weather. On Saturday a freshening breeze and falling barometer warned us to look out for squalls. In the evening the wind shifted to the south-west; all sail was then taken in and Sunday dawned with a furious gale right in our teeth. The sea was rising fast, and though well used to the ocean in all its moods; to see it cleft asunder by this huge ship as she drove at the rate of fourteen knots an hour into the storm, to watch

her splitting the great seas and sending the spray flying
like a continuous snowstorm from her bows, seemed to
me one of the grandest sights I had ever beheld.
For our ship was one—

"That neither cared for wind, nor hail, nor rain
Nor swelling waves, but through them did pass,
So proudly that she made them roar again."

At 10 A.M. she made one terrible plunge. Everything
passed out of sight in clouds of foam, and when her
bows rose once more we saw that the look-out bridge
had been smashed by the sea, and that the man in
it lay on the floor crushed down beneath the iron
rails. The engines were stopped to let the bo'sen's
crew go forward to lower him to the deck, and
he was taken to the surgery to have his wounds
dressed. Immediately the screw resumed work and
we drove on our headlong way into the storm. Two
hours later another huge billow loomed up ahead.
The engines were slowed. The great sea was split
and flung aside.

This time, however, the spirit of the storm had his
revenge; for as the big ship's bows rose out of the
foam we saw the bowsprit tremble and then plunge into
the sea on the port side. It had been snapped across
close to the stem. The engines were promptly stopped,
and when the ship's way through the water ceased,
the officers and men went forward to see what could
be done. The bowsprit was a hollow steel spar of about

four feet in diameter, it was hanging deep in the water, and being connected with the stem by solid iron bobstays there was no way of getting rid of it. The only thing to be done was to hoist it on deck. This operation took four hours, during which time we lay in the trough of the sea, with the full brunt of the storm on our broadside. Our ship, however, was marvellously steady; she did not roll much, but rose and fell with the greatest ease over the huge billows.

Next day we had steamed out of the cyclone, the summer sun shone out, and the afternoon was devoted to athletic sports on the promenade deck. The Fourth of July was duly celebrated by orations in the saloon, which was draped with flags in honour of the great day of American Independence.

On the evening of the 5th we were all on the look-out for land. The coast ought to have been visible, but a low-lying haze obscured it. Presently there loomed out of the fog an object like a rock rising above a strip of sandy beach, and an American friend standing near remarked, "That, sir, is the largest hotel in the world!" It certainly was a characteristic first glimpse of the country which every one knows "licks creation." Rockaway Hotel contains about 1700 bedrooms, and "the biggest hotel in the world" has smashed every company that, up to the present, has undertaken to run it.

As the sun was setting we threaded our way

through the lines of buoys, which mark the banks in the neighbourhood of Sandy Hook in New York Bay. Then we entered the harbour, between low headlands, clad in vivid green and fortified in old-fashioned style,[1] to the still waters inside, where we anchored for the night.

The western sky was glowing gold. A dark steel-blue gloom lay upon the harbour and city through which sparkled myriads of lights. High above the houses Brooklyn bridge was faintly visible, its bright lights glimmering like a chain of fire-flies in a spider's web, and nearer to us, Bartoldi's colossal statue of Liberty held up her electric torch, 300 feet above the waters of the harbour. The evening was calm and warm. Dancing went on on deck, and few regretted that we were late for the port doctor's visit, that functionary being off duty after sunset, and so must delay our landing till morning.

At 5.30 A.M. breakfast was served in the saloon. The *City of Rome* meanwhile steamed slowly up towards the wharves, entered her dock at 7 A.M., and we were then landed into a big shed where the custom-house ordeal had to be faced.

While waiting for our turn we were able to sym-

---

[1] Save and except the famous Zalinski pneumatic gun, throwing immense shells of dynamite, which on our homeward voyage we saw projecting like the long arm of a derrick from the fortress in mid-channel.

pathise with the misfortunes of others. One fellow-
passenger was so rigorously searched, that he was even
compelled to open a box of Cockle's pills, sealed up
in blue paper just as they had come from the chemist's.
Two Italians had to pay seven dollars duty on an
ivory crucifix in a velvet case. Our luggage was so
cumbrous that I expected all kinds of difficulties, but
the officers were most civil, and though puzzled a
little by our ice-axes &c., after a short consulta-
tion, they passed us without any trouble or payment
whatever. Entrusting our heavy baggage to an
Express man to deliver next day on board the Hudson river steamer, we stepped into a two-horse cab
and drove to our hotel about two miles distant. For
this drive we had to pay one dollar, which did
not seem excessive, considering all the warnings about
exorbitant cab fares we had received before leaving
home. As the temperature was over 90° in the
shade I was glad, after completing some matters of
business, to seek the breezy height of Brooklyn bridge.
I crossed it in the train and strolled back, looking
down on the city with its crowded wharves, the arms
of the sea which, with the Hudson river encircle it, and
the myriad white steamers crossing hither and thither,
making strange sounds with their whistles and fog-
horns, in all possible and impossible keys.

The broken outline of the city with its spires
and the colour of the houses, in which red pre-

dominated, contrasted well with the blue waters
of the bay, and the bright transparent atmosphere
all helped to make this one of the most interesting
panoramas to be seen anywhere in the world.

The evening we spent with an old college friend
and his family in their delightful home embosomed
in trees, near Orange, about seven miles from the city.
After the scorching glare of the streets it was delightful
to sit on the verandah in the cool night air and
watch the fireflies gliding through the shade.

As our intention was to leave New York *en
route* for Canada early next morning by the steamer
up the Hudson, we had to return by the last train to
Jersey city, and then in one of those wonderful
ferry steamers we regained New York.

The journey from New York to Montreal can be
made by various routes, but a glance at any map will
show you that a perfectly natural water-way exists with
quite insignificant breaks, by the Hudson river, Lake
George, and Lake Champlain. Before the days of railways
such a route was naturally of great importance, and
consequently from a historic point of view, as well as
from the beauty of its scenery, it is one of the most
interesting routes in America. Along these lakes
and rivers armies advanced and retreated in many wars,
and the battle-fields you pass in your journey—Saratoga,
Plattsburg, Ticonderoga, Crown Point and many more
recall both sad and glorious memories. Rival rail-

way lines now occupy both shores of the Hudson and make travelling expeditious and luxurious. We however determined to follow the course of the river and lakes in the big white steamers; those on the Hudson no doubt surpassing anything of the kind elsewhere, in the luxury of their fittings.

At 8 A.M. we went on board the *New York*, and found that our luggage had been duly delivered by the Express agent. Though I had been somewhat prepared for the magnificence of these river steamers, I felt, like the Queen of Sheba, that the half was not told me. The richly carpeted saloons extended from stem to stern, the abundance of sofas and artistic easy-chairs, the wood carving and general decorations were all magnificent, but the engine impressed me most of all. Large plate-glass windows, separating the saloons from the engine-room, enabled one to watch this splendid piece of mechanism. Every part of the engine and the rows of large spanners against the walls were burnished like silver and shone as I never saw iron do before. The motion too was unaccompanied by noise, and the engineer sat in an easy-chair on a carpeted floor.

The scenery on the Hudson is very beautiful. Bold precipitous hills, alternating with rich woodland, form the chief features of its shores. For many miles the Catskill mountains are visible. Numerous towns nestle along the margin of the river at the

foot of high hills, a continuous succession of parks
with comfortable country-houses give an old-world
aspect to the scene, and were it not for the lack of
old castles and vineyards, we might have thought we
were on the Rhine. In fact I think the natural scenery
is grander.

The huge icehouses, where ice cut on the river in
winter is stored for summer use, are very unsightly,
but the ice barges, as they pass in long lines down
the river, in the tow of steamers, form an interesting
feature of the Hudson, and give to it that air of busy
life which adds no small charm to the Rhine.

About 100 miles from New York, after steaming
some hours in sight of the Catskills, we reached Albany
and took the train for Saratoga. I do not know how
it was, but all day I could not get out of my head
my first trip to Switzerland. The Rhine, Wiesbaden,
the Swiss lakes and the snowy peaks all followed
each other in natural succession. Now the Rhine was
the Hudson, Wiesbaden was well represented by
its big American cousin Saratoga. But between
us and the snowfields lay, what has no parallel in
Europe—the boundless prairie. This is however going
a step too far for the present.

We spent Sunday at Congress Hall Hotel, Saratoga,
which supplies accommodation for a thousand guests.
A long imposing verandah, supported by Greek columns,
and a host of most obsequious but not-to-be-trifled-with

black waiters were the chief characteristics of our temporary abode. And as one had heard so much of the appalling fires in American hotels, it was no little comfort to find that every bedroom was supplied with a coil of strong manilla rope, spliced to an iron ring close inside the window and long enough to reach to the ground outside. It was comfortable to think that we need not be roasted alive; but we could not help speculating on what percentage of the thousand guests within, could slide even six feet down a rope, without letting go?

We went to church twice. At the evening service a judge from New Mexico, specially licensed by the bishop to preach, gave us a very good sermon concerning the building of churches on the frontiers of civilization in the far south-west. The band in the gardens played only sacred airs, as it was Sunday. Wishing to see what the country was like, H. and I took a three-mile walk. Far as ever we could see, the soil was of the lightest possible character, rounded gravel and sand predominating; the great amount of this kind of soil in the United States and Canada must be realised ere we can fairly estimate the population-supporting capabilities of the North American Continent.

On the principle of everything in America being "the biggest &c.," the Saratoga races are no exception. A race-meeting at home is considered long enough if it lasts two or three days. The Saratoga

C

races, with the "correct card" of which we were
furnished, continue for thirty days right on end.
We of course drank the waters at each of the
medicinal springs, and felt none the worse, and on
Monday morning were once more on the cars *en route*
for Lake George, which we reached at 9.30 A.M. through
beautiful wooded dells, bright with wild tiger-lilies.

Lake George is a thirty-five miles' length of perfect
loveliness; its thickly-wooded hills reminded me of
many Killarneys in one. As we entered bay after bay,
each had its charming little inn nestling amid rich
woods, its flotilla of skiffs and light canoes in which
pretty girls in pink and white muslin paddled to and
fro, or lay moored to tree branches while the occupants
lounged, or read, or sketched beneath the shade.

> " She was cargo and crew,
>   She was boatswain and skipper,
>   She was passenger too
>   Of the *Nutshell* canoe,
>   And the eyes were so blue
>   Of this sweet tiny tripper.
>   She was cargo and crew,
>   She was boatswain and skipper."

Life on Lake George in the glorious summer weather
seems quite worth living. How delightful it would be
to pause amongst those thousand islands, and to dream
with Fenimore Cooper of Mohicans, and the scenes
which these wooded hills once looked down on !

Times however were changed—the past might not

come back, the present might be a shade too charming, and our thoughts were already in the far wilder woods beyond the prairie. From the head of Lake George to Ticonderoga on Lake Champlain was half an hour by train. An interesting object in a scene, singularly wanting in natural beauty, were the ruins of Fort Ticonderoga, which crowned a flat, grass-covered promontory projecting into the lake. How many historical memories names such as these call up!

From the breastworks of this stronghold Abercrombie fell back, defeated by the French with a loss of 2,000 men, on July 16th, 1758, and only saved the survivors by a precipitate retreat down Lake George. The young and gallant Lord Howe, who fell on this fatal day, is commemorated in Westminster Abbey by a monument erected by the State of Massachusetts. A few months later however the French garrison capitulated to Abercrombie's successor in command of a larger force.

But the route we are on was also the theatre of still more momentous war—the great struggle between the American colonists and the British crown. Along this route Burgoyne fought his way by lake and by land against the Americans, till reaching the Hudson river he was defeated in the battle of Saratoga, and he and his army became prisoners of war. All the evening we spent steaming up Lake Champlain; past its busy manufacturing towns, which are now of more importance

than Ticonderoga of fateful memories. At Plattsburg,
also famous for its battle-field, we took the train
for Montreal, and arriving there in three hours,
a long but most delightful day was brought to a
conclusion.

Our first business in the morning was to look up
our luggage, which had come, booked through, from
New York, and was now in the bonded stores at the
Grand Trunk station. It had to be carted about two
miles to the Canadian Pacific Railway. Meanwhile we
went to pay our respects to Sir W. C. Van Horne, who
was then vice-president of the Canadian Pacific Railway. He was absent, but we were kindly received by
his secretary, Mr. Piers, who placed in our hands free
passes to Vancouver and back. A special pass entitled
us to make full use of the railway in the Selkirk range.

Mr. Piers also showed us a splendid collection of
photographs, taken along the line; so before we left
the head offices of the company we had learned
much about the regions we were so soon to see.
Our leave of absence from home being limited, and
not a moment could be wasted before we got at
our work, we were compelled to make this one
day do for Montreal; and as we hoped to gain much
further information at Ottawa, we left at 8.20 P.M.
Passing many farms along the line, which we could
see as long as the daylight lasted, we reached
Ottawa shortly after midnight.

My first thought after breakfast on the 11th was to visit Professor Macoun at the Geological Museum, but to my chagrin I found he had left for Nova Scotia. The young lady who acted as librarian however proved an able helper, and she provided me with all the maps and other information published by the Natural History Survey department. An hour was well spent in the Geological Museum, and then I walked to the government buildings and called on Major Deville, the Surveyor-General. He kindly showed me all the existing maps, gave me much valuable information, and allowed me to make a tracing of a small scale MS. map of the railway line through the Selkirks. This provided me with a starting-point for the survey we were about to make. Major Deville was also able to advise me in many practical details. He gave me a letter to Mr. Macarthur, the Assistant Surveyor, at present at work in the Rockies, and loaded me with " Reports of the Interior," out of which I picked much that was interesting on our journey.

The heat in the forenoon was most oppressive. About 2 P.M. it began to blow in gusts, and in half an hour there was a perfect hurricane, sending boards and everything movable flying in all directions, striking down telegraph wires, and causing general consternation; the temperature fell over 30° F. in four hours, and the evening turned out wet and bitterly cold. We were much impressed by the falls on the Ottawa river, flanked by great lumber mills which we visited when the storm had subsided ; but one day had to suffice for Montreal, so also must one day do for Ottawa, and at midnight we drove to the railway and took our places in the cars of the Pacific express, for our four days' journey to the Rocky Mountains.

CHAPTER III.

" What man would live coffined with brick and stone,
  Imprisoned from the influences of air,
  And cramped with selfish landmarks everywhere,
  When all before him stretches, furrowless and lone,
  The unmapped prairie none can fence or own ? "

LOWELL.

The Backwoods.—Winnipeg.—The Prairie.—Mementoes of the buffalo.
—Prairie Indians.—Calgary.—First view of the Rockies.

A COLD wet midnight is not the pleasantest time for
hunting up one's luggage on an open railway platform; so having secured a pocket full of brass checks,
representing the responsibility of the railway company
for an equal number of boxes and packs, we stepped
on board the nearest car. It was the colonists' sleeping
car, down which, between four rows of men, women,
and children, for the most part asleep, we made our
way to our own quarters. Selecting seats in the first-
class car, I went on to the "Sleeper" at the end of
the train, where I secured a luxurious upper berth,
and getting inside the curtains was snug between the

sheets, when the conductor's cry " All aboard," and the blast of the deep-toned whistle, more like a steamer's fog-horn than the shrill scream uttered by European locomotives, announced our departure from Ottawa for the Far West.

We passed through some cultivated country during the night, and found when leaving our berths in the morning that we had fairly entered upon the rugged back-woods region, abounding in moose and bears and innumerable rock-bound lakelets and trout streams, the happy hunting grounds of sportsmen; but except for "lumber," and in some places minerals, the land was worth little from a commercial point of view, and for the most part utterly irreclaimable. All day we rattled along through these regions, stopping every now and then at some little station, for our engine to take a drink, when many of the passengers alighted and gathered wild flowers. Some Indians with little papooses came on board at one station, and got out further on. We were in the back woods: but what a wreck! Occasionally we got glimpses of fine forests with stately trees, but for the most part nothing was to be seen but burning, smouldering, or charred trunks; the very soil seemed to be consumed, and the bare surfaces of ice-polished domes of ancient rocks appeared everywhere.

The smoke of the burning pines filled the air, and when night came on, the ruddy glare of forest fires made this wild region look still more weird. The railway engine is of course the chief culprit in this widespread devastation, and we hoped that the burnt-out belt did not extend many miles from the track.

Our second morning found us coasting along the north shore of Lake Superior, the track finding its way in wide sweeps and curves, round deep bays and bold headlands. The great inland ocean lay placid in the sunshine, only a very gentle swell made itself apparent in small breakers on the shore.

This portion of the Canadian Pacific Railway was the last constructed piece of the whole line, and was beset with the greatest engineering difficulties; it is also the most troublesome portion to keep open in winter, as the snow-storms blowing over the great lakes pile up heavy drifts in the numerous cuttings, which owing to the ruggedness of the route it was impossible to avoid.

The alternative but longer route, from Owen Sound to Port Arthur, by the fine steamers on the lakes, is adopted by many passengers going west, and at Port Arthur we were joined by this contingent, and till we reached Winnipeg, the train was crowded in consequence.

At Port Arthur we entered on the " Western division " of the Canadian Pacific Railway, which extends to Donald, on the Columbia, a distance of 1,454 miles. On our way to Winnipeg we traversed much of the route adopted by Lord Wolseley, in 1870, in his famous march on Fort Garry, as Winnipeg was then called, and saw some of his steamers, launches, and barges lying in a forlorn condition on the margin of a creek.

It was breakfast-time on our second morning when, after passing through much cultivated prairie land clothed with a green, luxuriant crop of young wheat, we reached Winnipeg. Here we breakfasted at the railway station instead of in the dining car, as was our wont; changed into another train, and then started for our long run of 900 miles over the prairie.

After quitting the rugged hills near Lake Superior it seemed as though we had left land behind us and were now crossing a great ocean. In truth we were crossing what had once been a wide expanse of sea, and that not very long ago, geologically speaking. For Canadian geologists tell us that in post-pleistocene times the old Laurentian and Huronian rocks of Eastern Canada were glacier-clad, and sent great ice-sheets down into a wide sea to the westward; that from these glaciers icebergs went adrift, as they now do from the Great Humboldt glacier in Greenland, bearing on their sides fragments of their Laurentian rocky bed : that they drifted across an ocean of water, dropping the rocks they bore as they floated along, where now waves the great ocean of grass and prairie flowers. In early times many such bergs were able to sail all the way to the Rocky Mountains, and the blocks of Laurentian rocks now found on their slopes

tell the story so plainly that he who runs may read.
In later times the ocean retreated to the eastward,
leaving the high prairie near Calgary as dry land.   For
ages after this the bergs still drifted westward, till they
grounded on that western shore which now marks a
great step in the prairie known as the Missouri Côteau.
Other ancient margins are visible where the ocean has
left landmarks of its retreat; and if the icebergs per-
formed no other great purpose in the economy of the
universe, they supplied numerous scratching-stones for
the buffaloes, who used them for ages till they became
quite polished.

So these boulders, scattered on the prairie, stand as
monuments of the past for three great facts: the existence of an ocean where now is land; the existence
of glacier ice where now the pine forests flourish; and
the existence, a few years since, of herds of buffalo
which roamed these plains in hundreds of thousands,
but like the ocean and the glaciers have vanished from
the scene.

The destruction of the buffaloes, sudden and bloody,
like man's awkward way of doing things, is a sad story,
but I suppose in some form it was necessary, for

   " Where man's convenience, health or safety interferes,
      His rights are paramount and must extinguish theirs."

The migrating of the herds was quite incompatible
with civilization and agriculture, for no fence could stay
their march or deflect them from their course.

For twenty years before their end came the hide hunters wrought destruction amongst them. The buffalo was a stupid beast, and if the hunter took ordinary care not to show his head, he might sit behind a bluff and fire away while his ammunition lasted without the herd moving off. A much more sportsmanlike way was to "run" the buffalo on a horse trained to run alongside the herd, while the rider planted bullets in the beasts he selected. Such famous shots as Dr. Carver could in a twenty minutes' run, kill or disable thirty or forty buffalo. In one season he is said to have killed 5000 to his own rifle.

The slaughter wrought in various ways caused the death-roll in some years to reach half a million.

Another reason for their destruction however brought the remnant that was left to its final doom. The Prairie Indians were troublesome ; a general massacre would not look well in the newspapers, so the best way to get rid of the Prairie Indian seemed to be to "wipe out" the buffalo on which he lived. So, valuable as their robes were, the sentence of death was pronounced by the American Government in 1879, and carried out in this way. Along the 49th parallel— the Canadian and United States boundary—the Americans stationed a cordon of soldiers, hunters, and Indians, with orders not to let a buffalo pass alive. And there, in that year, in sight of the Rocky Mountains, in the land of the Blackfeet, whose warriors,

from time immemorial used to fight for their rights
over the buffalo with their neighbours the Crees,
with the Flatheads, and Stonies, and Kootenays, who
traversed the Rocky Mountains to poach on their
preserves; 32,000 buffaloes were slaughtered in one
great hunt, and their bones left to whiten the plain.

Near Winnipeg, at Stony Mountain, a small herd of
buffalo have been preserved for cross-breeding with
European cattle, their chief value being the robes,
which now are so highly prized. Owing however to
the sterility of the offspring no great development is
possible in this direction.

Immediately west of Winnipeg we enter on the wild
prairie, which ere now would have been broken up by
the plough, were it not held by land speculators, who do
not cultivate, but just hold the land in hopes of gaining higher prices. As this great tract of country forms
a kind of shuttlecock between large capitalists, who
care nothing about farming, there is no knowing when
it will reach the farmer's hands and be turned to the
use for which it is so eminently fitted by nature. A
run of forty miles carried us through this unoccupied
belt. Then homesteads became frequent, with corn-
fields and pasture land, which would make any farmer's
heart leap for joy. Though the land looked as flat as
the ocean, the aneroid recorded a gentle increase in
altitude as we went westward.

Having now settled down somewhat, we began

to study our fellow passengers. Besides a detach-
ment of the Salvation Army, consisting of two men
and two "Hallelujah lasses," who sang hymns and
prayed at intervals during the day, we had a great
number of immigrants from distant lands; amongst
them about a score of Icelanders, the women very re-
fined-looking, in full costume decked with silver
ornaments and black lace, the men rough and hardy.
They could speak no English, and at Brandon, the
largest prairie town in Manitoba, they left us and were
conducted to the sheds prepared by Government for
immigrants until they could be settled on their allotted
land. About 6 P.M. we passed out of Manitoba into
the Western territories, and when the sun set we were
in an ocean of wild prairie grass. A sedgy lake
teeming with wild fowl; innumerable prairie dogs, like
elongated guinea-pigs, nibbling the seed off the grass,
or standing erect at their burrows; and some lodges
of Cree Indians made up the landscape.

During the night, we stopped at Regina, the capital
of the province of Assiniboia, where a branch of the
Canadian Pacific Railway turns off to the northward.
Here are the head-quarters of the Mounted Police, a
fine military-looking, red-coated body of cavalry, to
whose care the peace of the territories is intrusted.
Here too meteorological records are kept, and are
specially interesting, as in this region, the centre of
the plains, we find great extremes of summer heat and

winter cold. In the latest statistics in my possession
I find that for January and February the mean temperature at Regina was 15° below zero F., and the
lowest reading recorded was − 52°, while in July the
mean temperature was + 62°. The rainfall for the
same year was under twenty inches. The dryness of the
air and absence of wind makes heat and cold far more
tolerable than one might think from the mere readings
of the thermometer; but when the cold comes to mean
seventy or eighty degrees of frost it is rather too severe
to be either safe or pleasant.

· When morning dawned and the sun rose, the view
was exactly like what it had been the night before, a
rolling prairie with a horizon-line like that of the
ocean. Crossing and recrossing it in all directions
were the tracks of the buffalo, and here and there
might be seen the skull or bones of some of these fine
animals, bleaching under the summer sun. At some
of the tanks where we stopped to water the engine,
heaps of skulls and bones had been collected for ex-
portation to the sugar refineries or manure works of
civilization.

Indians came to the train selling the horns they had
picked up, but, as these were polished and thus
robbed of all character, they did not seem attractive.
On our way home I secured one of the whitened
skulls, with its rough weather-beaten horns.

Besides the tracks of the buffalo, we saw many of

their wallowing holes, dotting the plain. Here having dug a slight depression with their horns, into which the water trickled, they rolled and covered themselves with mud. All day long we passed an unending series of grass-covered swells as in a panorama; its very sameness seemed so wonderful and so strange that I could never tire of looking at it.

In the afternoon the land appeared to get more sandy and barren, the sage bush was more frequent, and the profusion of sunflowers marvellous—they spread out everywhere in sheets of gold, but along the railway track they were richest of all. The margin of the line was like a ribbon of intense yellow far as the eye could see.

At last at 4 P.M. a change came over the scene. Passing through some low, barren-looking hills, the line descended into the valley of the Saskatchewan river and we stopped for twenty-five minutes at Medicine Hat station. Here our tired-out locomotive was changed for another, the wheels of the cars were overhauled, and away we went again to the westward. As we crept up the incline beyond the river, an Indian on his pony raced the train, and seemed quite elated with himself at getting ahead of us, crossing the track in front of the engine, and riding slowly down the opposite side of the train.

About sunset we saw a few antelopes and coyotes, the latter evidently hoping to pick up a stray prairie dog for supper; some large hawks sailed about with no doubt a similar object in view.

Our elevation above the sea had now increased considerably, the general level of this part of the prairie being about 2,500 feet. The air was bracing and crystal-clear. In summer at any rate no climate could be more delightful; and strange as it may appear, from this on to the mountains, though increasing gradually in height, this region possesses the most open climate in winter of any portion of the Dominion east of the Rocky Mountains.

The shades of evening soon closed in, and it was long after midnight when we reached Calgary. As it was our intention to break the journey here, in order to buy provisions for our mountain expedition, and to see some friends, we left the train and went to a hotel. It was a grateful rest after days of perpetual motion and rattle; and as no West-bound train would pass until the same hour next night, we took our time at breakfast, and then went to call on the worthy parson, who conducted us to the best store for purchasing bacon, flour, biscuits, tinned meats, tea, sugar and all the *et ceteras* which seem necessaries of life. How strangely all plans of travel finally gravitate to the commissariat! If only some one would kindly guarantee the commissariat, the summit of Mount Everest and the South Pole might easily be reached.

I was much interested in hearing something of the

D

Indians, who occupy three large camps near Calgary. They belong to different tribes; the Bloods and the Blackfeet speaking one language, and the Sarcees another.

At dinner at our friend's house we had the good fortune to meet the mission teacher of the S.P.G., who resides with the Sarcees, and was formerly for some time in the much larger Blood camp. As these camps were too far off to reach them in our limited time, we walked about four miles on the prairie and saw some lodges of the Blackfeet. All these prairie Indians adhere to their native dress, with the addition very commonly of a slouched hat, which does not correspond somehow with the rest of their gear—a kind of tunic, leggings with coloured fringes all down the back of the leg and a bright-coloured blanket. Since the land has been taken up by Government they are supposed to live on reserves, and for the most part the law is adhered to: on the reserve is a Government agent who issues rations to them, which, with other items, form portion of the stipulated payment for the landed rights handed over by them to the Government. They breed horses, learn agriculture, and go on the hunting path at stated times of the year as formerly. Though they still endure great tortures in the ordeal which admits them to the number of the braves, their wars we hope are over for ever.

We ought I suppose to feel thankful for such a

change.   How strange then does it seem that sadness
is the feeling which involuntarily seems to arise
instead !   The fact is, the Indian was like a fine wild
animal in the ideal we formed of him from story-
books.   Now he is dropping into a very ordinary
piece of humanity.   We must in truth be thankful
that civilization, and we hope in some cases real
Christian feeling, has made their ferocity, their love
of torturing captives, and some of their more terrible
vices impossible.   Well for them if the meaner vices
which skulk under a varnish of civilization do not
take their place, and leave the last state of the man
worse than the first !   The Indians living on the
prairies of the Dominion and in British Columbia
now number about 100,000, but an accurate census of
them would I think be impossible.

The settled Indians of the Eastern provinces have
been admitted to the franchise.   The half-wild Indians
of the prairie are yet outside these rights of citizenship.
They are essentially nomadic, and their camps show
the necessity for constant moves.   Around their lodges
bones and skins and offal of all kinds accumulate and
their only plan of improving their sanitary condition is
to strike camp and move on.   Every acre of the plain
near Calgary appears to have been occupied in this way.
It seems as if it were yet a long way off before such a
race of people could thrive in settled towns.

The interest which attaches to these poor Indians has

D 2

delayed me from saying one word of the view. But
what a view! worth coming all the way from England
to see. In a deep cutting the Bow river wound its
way towards the mighty Saskatchewan : the great highway of the North West, before the Canadian Pacific
Railway revolutionised that region. North, south, and
east lay the wide swells of golden-grassed prairie, but
to the westward the Rocky Mountains were in sight for
nigh a hundred miles. They rose like a great purple
rampart, jagged and peaked in outline, above the ocean
of grass. Glaciers and snow fields glinted in the sunshine ; deep valleys suggested rivers and passes ; the
distance was too great to make out the details, the
sharp outline of the summits melted downwards into
blue atmosphere, as the lower portions of the ranges
met the golden yellow of the prairie : the contrast was
superb.

CHAPTER IV.

" Where living things, and things inanimate,
Do speak, at Heaven's command, to eye and ear,
And speak to social reason's inner sense,
With inarticulate language."

WORDSWORTH.

Descriptive.—Geology.—Forests.—Animals.—Early explorers.

As we are about to leave the plains across which we have travelled for more than a thousand miles, most impressive as they were to us in their vastness; and are about to enter the defiles of the Rocky Mountains, it seems well that we should try and form some clear idea of the nature and chief physical aspects of this great mountain region.

It begins abruptly, as I have described in the first mountain range, the Rocky Mountains proper, rising as a great rampart and bounding the western margin of the prairie. The foot-hill region, where the prairie, though cleft by river valleys and rapidly rising into tilted slopes, retains its character for some time in benches and plateaus, forms a belt only fifteen miles wide. To

the westward of this first range, the high peaks of
which average about 11,000 feet above the sea, we
come to a deep depression filled by the upper portion
of the Columbia river, which, rising far to the southward of the track followed by the railway, flows northward, parallel to the Rocky Mountains, for 170 miles.
Then, having been joined by Canoe river from the
glaciers of Mounts Lyell, Hooker, and Brown, which
mountains are probably from 10,000 to 12,000 feet high,
it sweeps round in a wide bend and flows southward,
parallel to its former course. Passing through the
Arrow lakes it is joined by the Kootenay. This large
tributary near its source is separated by only one and
a half miles from the head waters of the Columbia:
after flowing southward in a wide bend and passing
through the great Kootenay lake it joins the Columbia,
which has now assumed the dimensions of a mighty
river. The Columbia now crosses the United States
frontier, and after a total course of 1,400 miles, finds
its way through Washington territory to the Pacific.

Entering this mountain region from the eastward
by the Canadian Pacific Railway, we cross the first
mountain range at Hector pass, which, involving an
ascent from the prairie of only 2,500 feet, descends
4,360 feet by steep gradients to the Columbia valley.
Making a first crossing of the river at Donald, the
track then ascends the Selkirks, as those mountains
are called which occupy the whole region inclosed by

the great loop of the Columbia and Kootenay rivers. At Rogers pass, 4,275 feet high, the railway crosses this range. Most picturesque snow-clad peaks tower up 6,000 feet above the track, and white glaciers are visible above the rich dark pine forest. Descending once more into the valley of the Columbia, in the western portion of its course, the line crosses the river for a second time, and then ascending the Gold range, and skirting for many miles the lovely Shushwap lake, it makes its way into the valley of the Thompson river.

Following the Thompson through its grand cañons to its junction with the Frazer, the railway goes along by the Frazer through magnificent gorges to the flat forest land near the sea. It will be seen then, that on leaving the prairie we bid adieu to the plains and for several hundred miles traverse a succession of mountain ranges; the Rockies, the Selkirks, the Gold range and the Cascades, being, roughly speaking, parallel. We enter a region of vast ravines and wide valleys, whose sides, when not bare rock precipices, are clad in sombre forests; through which wild mountain torrents rush from glacier sources to placid lakes, where after resting for a while and reflecting the hoary cedars and hemlocks, they issue forth as great rivers and with swift current hurry on to lose themselves in the Pacific.

The geology of Europe, where that science was first studied, under most difficult and complicated circumstances, forms a striking contrast to that of the New World, where geological structure exhibits itself in grand simplicity.

Many centuries of investigation passed before men recognized that the principal rock formations had been laid down as sediment at the bottom of ancient seas. Let any one look upon the walls of the great cañons of the Colorado, or even at photographs of them, and he will find illustrations which at once force this conclusion on his notice. The fossiliferous deposits they contain confirm the fact and forbid any other possible explanation. The strata through which the Colorado has cut its grand section are the same which at an almost dead level cover the central part of the whole North American continent. In Colorado such a section is possible, owing to the strata having been gently upheaved above the sea-level; but so gentle is the upheaval that the horizontal character of the deposits has in no way been interfered with. From these great sections we learn that a large part of the American continent, now forming plains and plateaus, was gradually and continuously subsiding from carboniferous times to the end of the cretaceous age. The mud deposits of this long period, even in their present consolidated state, attain a thickness of about 15,000 feet, and since cretaceous times a gradual upheaval has been going on.

In some places the land has never since then

been covered by sea, though it may have borne fresh-water lakes, the deposits of which have been distinctly recognized.

In the prairie region where the elevation has not been so great, the sea held its sway until more recent times. The great carboniferous ocean, in the portion of the continent crossed by the Canadian Pacific Railway, extended over the area now occupied by the Rocky Mountains. From our investigations in the Selkirks, and from the examination of our rock specimens by Professor Bonney since our return, I would conclude that the Selkirk peaks looked down upon that carboniferous sea; for the archæan rocks of the Selkirks and the more ancient character of all the strata of their higher ridges seem to point to their having been land before the Rockies of this region had risen from the sea.

The elevation of the more eastern range (*i.e.* the Rockies) commenced before, and continued subsequent to the deposition of sediment in the cretaceous ocean. This ocean seems to have been bounded at the foot of the Rocky Mountain region by extensive swamps, the deposits of which are now being worked in the Bow river valley and farther out in the foot-hill region as valuable coal beds.

The submergence of the prairie region in glacial times has already been referred to. A Mediterranean sea, according to Sir J. Dawson, seems then to have existed, open to the southward, across which icebergs drifted, from the Laurentian land of Eastern Canada to the shores of the Rockies.

In the upheaval of the Selkirk range the structure of the continent has been disturbed to its lowest archæan foundation. Cambrian rocks seem to be the oldest in the Rockies. In both ranges there has been a violent crumbling and contortion of the strata, causing much metamorphism in the process.

The rocks of most constant occurrence in the Selkirks are hard white and greenish, fine-grained quartzites, and silky corrugated mica schist. In the Rockies quartzites are also of frequent occurrence, and the Ottertail range forming a beautiful group of Alpine peaks, is composed of an intrusive mass of syenite, a rare phenomenon in this region.

The higher portions of the three parallel ranges, the Rockies, the Selkirks, and the Gold mountains, are glacier-clad. So far as we could judge, the lowest termination of the glaciers is at about 4,000 feet above the sea in the Selkirks, and 1,000 feet higher in the Rockies; the difference must be accounted for by the greater amount of moisture deposited on the former range.

The greater moisture of the Selkirk climate is abundantly illustrated by the vegetation.

The forests in the Rockies are comparatively open, so that pack-horses can be taken nearly everywhere through them. Those of the Selkirks are so luxuriant as to be well nigh impenetrable.

About twenty species of Coniferæ occur in the forests of this mountain region. All the pines which flourish in the Rockies grow luxuriantly in the Selkirks, but in the Rockies such splendid trees as the western white cedar,[1] and western hemlock[2] are not to be found except it may be sparingly on their western slopes; while in the Selkirks they form a most important portion of the forest, attaining sometimes a height of over 150 feet, and the trunks of the former tree sometimes being ten feet in diameter.

Some species, such as the fine Douglas fir,[3] prefer the valleys; others, like the balsam,[4] flourish close to the snow-line. Some species form extensive groves all by themselves; as a rule, however, many varieties are mixed together, and Engelmann's spruce[5] may be considered, of all others, the most characteristic. In the Selkirks the undergrowth makes the forests most difficult to travel through. Besides such low-growing Coniferæ as the western yew,[6] much of the scrub consists of white-flowered rhododendrons.[7] Blueberry bushes[8] grow everywhere; their delicious fruit is a most important help to the traveller's slender commissariat, and it has saved many a prospector from

[1] *Thuya gigantea.*
[2] *Tsuga Mertensiana.*
[3] *Pseudotsuga Douglasii.*
[4] *Abies subalpina.*
[5] *Picea Engelmanni.*
[6] *Taxus brevifolia.*
[7] *R. albiflora.*
[8] *Vaccinium Myrtillus,* &c.

starvation. In the damper portions of the forest the
devil's club [1] grows luxuriantly. It has large pale-
green, palmate leaves and tufts of coral-red berries,
and attains a height of from four to six feet. But
its stem and leaves are so beset with long slender
thorns that it cannot be touched with safety, as the
thorns are said to produce festering sores. It is bad
enough, but probably the worst obstacle of all to the
traveller is the alder scrub, which forms a dense
jungle of slender stems, the lower parts of which
from being pressed down by snow when young, grow
parallel to the ground and then curve upwards to the
vertical. Alders [2] thrive in the upper parts of the
valleys near to the glaciers, and wherever else the pine
forest has been prevented, by constantly recurring
avalanches, from establishing itself.

When the configuration of the ground is favourable
there is often above the forest line an area covered
with grass, heaths of various species, and interesting
Alpine flowers, to which I shall have many occasions
to allude. Comparing this Alpine flora with that of
Switzerland we find many species identical, and the
colours of the flowers seem quite as varied and brilliant.

In this respect I observed a distinct contrast to the
Alpine flora of New Zealand, where blue and red seemed
absent, pale tints predominated, and the majority of
the flowers were either white or yellow.

_____

[1] *Fatsia horrida.*   [2] *Alnus rhombifolia* and *virescens.*

Though the flora of the Selkirks seems richer than that of the Rockies, animal life is not nearly so varied. Bighorn and wild goat are the most characteristic animals of the Alpine ranges of the Rockies. The former does not occur at all in the Selkirks.

The wild goat [1] is a fine creature, though not a true goat, but a goat-antelope like the chamois. Unlike the chamois, however, it is covered with long white wool. The hoofs are very large, enabling it to walk on soft snow, and its tracks are consequently about the same size as those of a yearling calf. Where we met with it, it was comparatively tame from never having been disturbed, and was inclined rather to seek our company than to shun it.

I cannot now go into the details concerning the deer found in the lower ranges, nor the numberless small animals of the marten and squirrel families inhabiting the forests. I must however say a few words about the bears. Black bears are very numerous, and we saw skins of the cinnamon bear, and the silvertip, as the variety of grizzly inhabiting the Selkirks is called. We constantly found ourselves on the fresh tracks of bears. They were often close to our camp. We measured their footprints in soft mud, but we never got a shot at one. The fact seems to be, that if you want to shoot bears in summer, you must have dogs to hunt them up and bring them

[1] *Aplocerus montanus.*

to bay. Rough Irish terriers are as good for this work as any. The skin is worth little in the summer, so the best time for bear hunting is when the winter fur is growing, just before they go into winter quarters, or else when they are waking up in the spring. Bears will almost invariably shun the presence of man, so you are perfectly safe from unprovoked attack; there is danger however, in meeting a she-bear with cubs. A wounded bear is certain to attack, so when you fire you must know what you mean to do next. A Winchester rifle, the weapon in general use, gives you every advantage if you have time to use it. An axiom in bear-hunting therefore is—never fire at a bear above you on the mountain side, for he can charge so fast downhill that he may give you little time for a fatal shot. I could only hear of one solitary case of a bear having attacked a man, without provocation. The man was lying on his face, drinking at a spring, when the bear jumped on him. His companions fortunately shot the beast at once, and saved him from anything worse than a bad scratching.

On the lakes and rivers of the low ground throughout this region there is an abundance of geese, ducks, and numerous other valuable birds. In the high ranges bird life is decidedly scarce. Grouse of three species are to be met with, one, the "fool hen" inhabiting the forest region, being exceedingly tame and stupid, as its name implies. The blue grouse of the upper slopes

formed an excellent addition to our camp meals. On
the Columbia river we saw many white-headed eagles
and fish hawks. Among the Alpine heights I do not
remember having once seen a rapacious bird.

If we were to judge of the climate of the Selkirks
from our own experience only, I would conclude that
such a splendid Alpine climate could not be surpassed
by any other in the world. And though we were
specially fortunate, fine weather in the summer is the
rule. The annual rainfall in the Selkirks is, as might
be expected, much greater than in the Rockies.
There is therefore more snow and the richer growth
of vegetation to which I have already referred.
When the winds from the Pacific reach the prairies to
the eastward, much of their moisture has been deposited
on the various ranges they have traversed, and as a
consequence, their temperature has risen, so that
these warm "Chinook" winds, as they are called,
form a most characteristic feature of the whole region.
The winter climate of the high prairies of Alberta is
so much under this influence, that cattle can remain
out all the winter, and snow never lies for any length
of time.

In the month of January, under the influence of a
northerly wind, the temperature of these high plains may
be lowered to a maximum of $-19°$ F. for a day or two,
but then a Chinook wind sets in, and within a week
we find readings of $45°$ and $50°$. At the eastern foot

of the mountain region the influence of this wind is at its maximum, but passing westward, we hear of the Chinook wind bringing a warm temperature in the Columbia valley and even on the western slopes of the Selkirks. The great snow slides of the early spring are set down to this influence. The moisture deposited on the ranges nearer to the West Coast, must be the cause of this early rise in temperature. Most mountain countries experience winds of this nature; the Föhn of Switzerland, and the hot winds of the Canterbury plains of New Zealand are exactly similar phenomena.

Now that the Canadian Pacific Railway has penetrated this region, its phenomena can be studied year after year, and it is hard to realise that, till five short years ago, it was one of the almost inaccessible wildernesses of the great North West.

The exploration of these western wildernesses of mountain and forest was for many years left to the enterprise of the great Fur Companies; the Hudson's Bay and the North West Companies being the chief rivals in the field.

They established forts in the Indian country and traded with the Crees, the Blackfeet and other tribes, but the jealousy, which existed between the rival companies was more intense than we now-a-days can well understand. The agents in these forts were British subjects, for the most part Scotchmen, but those belonging to the one company would sometimes, with a whole horde of whooping Indians at their back go off on the war path, attack the other fort, burn it down and return triumphant with scalps. Then would follow reprisals backed by other tribes of Indians.

In some portions of the territories matters were on a better basis. For example, where one company was satisfied to hire professional bullies and locate them in a hut near the rival store, to terrorise would-be traders from coming near it. In other cases the agents were content with swindling each other decently. But that the above state of affairs could exist at all is a strange fact in the history of civilization.

One of the earliest explorers of the region which we are now describing was David Thompson, who went out from England in 1789 in the employment of the Hudson's Bay Company. In 1797 he left it to join the rival North-West Company, under which he now undertook a most wonderful series of ex- plorations, extending from the Mississippi to the great Slave Lake and from Lake Superior to the Pacific Ocean. In 1800 we find him exploring the Rocky Mountains by the valley of the Bow river, and in 1808 he has established a fort on the Columbia lakes and is trading with the Kootenay Indians.

Almost all the early pioneering expeditions in these North-West territories and British Columbia were led by Scotchmen, aided by French-Canadian *voyageurs*,

E

and nobly the work was done. Not only did Scotland send out fur traders, but also men like David Douglas, after whom the fine Douglas fir takes its name. He was a young Scotchman who came out under the auspices of the Royal Horticultural Society of London in 1824, and for ten years explored the great forests of the West, studying the birds, beasts, and plants. "To the botanical vocabulary of the time David Douglas added the names of over one thousand plants. Thus this devotee of birds and plants wandered among the forests of America, his pack on his back, and a shaggy terrier at his heels." [1] He was a man of plain speech, but on one occasion when staying at Fort Kamloops very nearly lost his life for remarking to the brother Scot in command of the fort that "the Hudson's Bay Company had not an officer with a soul above a beaver skin."

The pass across the Rocky Mountains in most frequent use for many years by the fur traders was that known as the Athabasca pass, connecting the northern bend of the Columbia with the eastern river basins.

The pass used by the Canadian Pacific Railway was discovered by Dr. Hector, who accompanied Captain Palliser's exploring expedition in 1858. It was called by Dr. Hector the Kicking Horse pass, after the river, on the banks of which, he received a kick from a horse, which disabled him. As his discovery of this pass solved the

[1] Bancroft, *Hist. of N. W. Coast.*

question as to a practicable route across the mountains for a railway, I give Dr. Hector's own words. He had crossed the range to the westward, and was now seeking a way back over the range. "August 29th. Reached the mouth of a large tributary, to north-west. . . . Here I received a severe kick in the chest from my horse, rendering me senseless, and disabling me for some time. My recovery might have been much more tedious than it was, but for the fact that we were now starving, and I found it absolutely necessary to push on after two days. On 31st August we struck up the valley of the Kicking Horse river, travelling as fast as we could get our jaded horses to go, and as I could bear the motion, and on the 2nd of September reached the height of land." [1]

This discovery of Dr. Hector's was thus commented on by Captain Palliser, in his report to the British Government. "In that pass Dr. Hector had observed a peculiarity which distinguishes it from the others we had examined, viz. the absence of any abrupt step at the commencement of the descent to the west. This led him to report very favourably upon the facilities offered by this pass for the construction of a waggon road, and even that the project of a railroad by this route across the Rocky Mountains might be reasonably entertained."

[1] "Height of land" a frequently-used term, meaning "Divide" or "Watershed."

No railway had then been made to link ocean with ocean, and it seems most suitable that now after thirty years the pass should for the future bear Dr Hector's own name, instead of one recalling the memory of his misfortune.

Time went on, other lines crossed the continent, and Great Britain could not be behindhand in enterprises of this nature. Many passes through the Rockies were explored. The Kicking Horse pass could not be beaten, but then there was the Selkirk range beyond. Mr. Walter Moberly had here been the chief explorer, but the railway difficulty was at last solved in 1883 by Major Rogers discovering the pass bearing his name, and also a pass through the Gold range to the westward, the last link necessary to connect ocean with ocean. Thus was Sir James Hector's suggestion of thirty years ago fulfilled, and the Canadian Pacific Railway became an accomplished fact.

CHAPTER V.

There its dusky blue clusters
The aconite spreads,
There the pines slope, the cloud stripes
Hung soft on their heads.
No life! but at moments,
The mountain bees hum,
I come, O ye mountains!
Ye pine woods, I come!

MATTHEW ARNOLD.

Leave Calgary.—Ascend the Rockies.—Steep descent.—Cross the
Columbia.—In the Selkirks.—The trestle-bridges and snow sheds.
—The great Illecellewaet glacier.

OUR walk on the prairie had proved so delightfully refreshing, that when the Pacific Express arrived soon
after midnight, and we had taken our places in the
cars, we sank quickly into dream-land, and did not
wake up till the train was far into the bosom of the
mountains. On awakening, the outlook seemed very
beautiful. The Bow river, along the margin of which
we ascended, flashed and sparkled in the morning sunshine. Wreaths of filmy golden mist hung around
the sombre pine forest, while above all, the mountains

rose on either hand in beetling cliffs and snowy summits.

It was a glorious morning, and to a lover of mountain scenery this first near glimpse of the Rocky Mountains, resplendent in the golden light of the rising sun, was one of those experiences in life never to be forgotten. During our time of blissful unconsciousness we had passed Banff, where a very fine hotel has been built by the Canadian Pacific Railway, and were now drawing near to Castle Mountain station.

Here we paused at the water tank, and Mr. McArthur, Assistant Government Surveyor, to whom we had been commended by his chief at Ottawa, came on board the train, from his camp close by. He, with his assistant, was *en route* for the summit of the pass, from which he intended making a mountain ascent for surveying purposes. As he had been for two or three seasons engaged in the survey of this portion of the Rockies, he could give us much interesting information. As we rattled along, he pointed out the various mountains of importance, and said he would like to join us later on in an attempt to ascend Mount Lefroy (11,658 feet), the highest measured mountain in the portion of the Rockies in British territory. Though our special work lay in the Selkirk range, the prospect of exploring some of the peaks of this eastern range seemed most agreeable, so we settled that morning before we parted, that I should leave enough time for this expedition on

our return journey, and give him due notice when to
expect us. Assuring us that we might trust all camp
arrangements to him, he and his companion left us at
Hector pass for their mountain climb, and bidding
them *bon voyage*, we entered upon the most critical
portion of the whole line, the terribly steep descent
towards the Columbia valley.

At the summit of the pass a huge locomotive with
ten driving wheels, and weighing one hundred and
seventeen tons, was attached to the back of the
train, and trusting to its restraining power and to that
of eight extra men, who came on board the train to
help at screwing down the brakes attached to each car,
we started at a cautiously slow pace down a gradient
of one in twenty-three. The Wapta river, in whose
company we had to make our way to the Columbia, on
issuing from its lake on the summit of the pass, plunges
down in a series of cascades, descending 1,100 feet in
five miles. The railway track, being unable to descend
in this precipitate manner, clings to the steep precipices
of the mountain side and consequently is soon left
high above the valley. Across trestle-bridges spanning
deep ravines, and round sharp curves, we wound our
way, getting views from the windows of the train, or
better still, from the platform at the end of the car,
which were sufficiently startling. Pinned on to the
face of the precipice, trusting in many places to
elaborate scaffoldings of pine trunks, built up from

what seemed perilously insecure foundations, occasionally
resting on mere notches in the rocky walls, the track
winds its way downwards to Field station at the foot of
Mount Stephen, and the level of the Wapta is once
more reached.

Here the big locomotive left us to await the
arrival of the East-bound train, which with mighty
puffings it had to shove up the steep incline to the
summit. As it is impossible to take the heavy dining-
cars through the mountains, three little inns have
been built by the Railway Company, one at Field, one
the "Glacier House," at the summit of the Selkirks,
and one at North Bend on the Frazer river. Each
of these has its refreshment room where excellent
meals are served, and here at Field all the passengers
betook themselves to breakfast. Porridge, broiled
salmon, beefsteak, omelette, and mountain air seemed
to go well together, and every one felt more happy
when the conductor's cry of "All aboard," and the
blast of the whistle resounding from the overhanging
cliffs warned us to resume our seats and continue on
our way.

The views up lateral valleys to glacier-clad peaks,
every moment attracted our attention, but our
grandest mountain view was that of the Ottertail
range, to the south of the track. The peaks of this
range, composed of intrusive syenite, tower up in
sharp aiguilles, the hollows being filled with glacier ice.

I commend them to all lovers of Alpine scenery for
their grand outlines, and few things did we put
aside with greater regret than a half-formed plan
to break our return journey at Leancoil station
and see more of the Ottertails.

Before the thoughts of scrambles among these
grand peaks had given our minds a chance to think
of anything else, we had shot into the cañon of the
Wapta. The river roared between steep rock walls,
almost splashing the train, which now disputed
possession of the outlet which it had taken the river
ages to cut ; and as we rushed along, we seemed as
if racing the torrent, in its headlong career towards
the Columbia. Deeper and deeper we went from the
light of day, the noise of the torrent and the rattle
of the train blended into one loud roar, which echoed
and re-echoed from the inclosing precipices rendering
our voices inaudible. Scarcely a vestige of vegetation
could find foothold in the dark gorge. Nothing
but river and railway, smoke, spray, and steam.
Suddenly the light became stronger, the great roar
was hushed, and we emerged from the savage
Wapta cañon into the wide smiling valley of the
Columbia.

We had descended from the regions of snow and
ice into an almost tropical climate. The heat was
intense, and to add to the resemblance of the tropics,
mosquitoes set to work upon us with the greatest

avidity. A few hundred yards from the cañon the train came to a stand at Golden City.

This place consists of two inns, a store and a number of log houses. It owes its importance to its being the lower termination of the navigable portion of the upper Columbia, into which river the Wapta pours its turbid waters. The houses stand on low ground which, with a wide area of gravel sparsely covered with bushes extending to the Columbia, formed at some time the bed of the Wapta, and will if we mistake not ere long be claimed by the river again. The natural action of such a torrent which for forty miles comes leaping and dashing down the mountain side, fully charged with *débris*, is to deposit that *débris* very rapidly in this its first period of comparative rest. When the river finds a depression it immediately seizes on it as a channel. The silting up of its new bed commences at once, so that soon what was a depression ceases to be one, and the river bursts off in some new direction seeking fresh channels. The inhabitants of Golden City fondly imagine that a little raising of the banks will keep the river to its course. No, not if its banks were built up till they topped the roofs of the present houses, could such a river be restrained for more than a very limited period. Those who would "build for aye" had better betake themselves and their houses to the high ground above the railway.

In 1884 Golden City bore such an evil reputation
that H. and his companions when passing on their
journey, had given it a wide berth, and camped further
down the Columbia. These wild days of its youth
are over. We spent two safe and comfortable nights
in Golden City when *en route* for the Columbia
lakes.

A twenty miles' run down the valley of the Columbia
gave us ample time to study the flanks of the Selkirk
range, which rose in great swells of unbroken forest on
the farther side of the river. The higher peaks were
not yet visible, and it was only here and there, above
mountains of pine forest, that we got glimpses of
snowy peaks. When we stopped to change engines at
Donald, we made the acquaintance of Mr. Marpole,
superintendent of the Selkirk section of the railway,
and to whom we were afterwards much indebted for
the kind help he afforded us in our work. From
Donald we made our first crossing of the majestic
Columbia, which swept steadily on in swirling eddies,
with broken reflections here and there of the dark
cedars and firs. Following the river for a few
miles further, until it broke into foaming rapids, we
turned sharply into the ravine of Beaver Creek, and
commenced our ascent of the Selkirks. The magnificence of the forest was wonderful. It was composed of
cedar and spruce and the huge hemlock, which latter
predominated as we increased our elevation.

Higher and higher we crept along the mountain side, gradually leaving the Beaver far below. Deep ravines cleft by glacier streams, foaming down, half choked with fallen logs, were spanned by lofty trestle-bridges. One of these which we crept slowly over was Mountain Creek bridge, and we went still more slowly over those portions of the track which were laid in shallow cuttings in the loose *débris* which lay at a high angle on the mountain side. We could see showers of gravel shaking down as the train passed. Then we came to the most wonderful bridge of all, spanning Stony Creek, at the prodigious height of 295 feet above the torrent, chiefly supported by one tall pillar of trestle work, rising straight up from the bottom of the ravine, and a smaller one which is secured to the sloping side of the chasm.

The one thing which spoiled the prospect, as we looked from the platform of the car, was the charred and burnt forest. In some places the timber was destroyed all the way from the railway down to the river and up to the mountain summits beyond. There was no sign of fire at present, for the weather up to this had been wet, but grey and black branchless trunks stood up as ghastly records of the fires which had raged during the previous autumn. Some of these trunks were of vast size, eight to ten feet in diameter at the butt, and probably over 150 feet high. The action of the fire was peculiar: in most places the bark

" A series of snow-sheds."—P. 61.

was burnt off, and then the fire seemed to eat its way
up through the centre, hollowing out the whole trunk.

Later on we saw only too much of these destroying
fires ; now the air was clear of smoke, and as the higher
peaks came into view every crag and crevice was easily
discernible.

At a station called Bear Creek, where the valley
forked, we left the Beaver, and following the valley of
the Bear the train had to pass through a series of
snow-sheds. On emerging from these we found ourselves in the wonderful defile between Mounts Tupper
and Macdonald, whose great precipices rose so vertically, that we could only see to the top of
the precipice opposite by leaning far out of the
windows. As we crept round the base of Mount
Tupper and entered Rogers pass, our prospect was
sadly interfered with by snow-shedding. We no sooner
were out of one snow-shed, and had got merely a
glimpse of the magnificent scenery through which we
were passing, when we went into another. These
sheds are built of massive timber-work, all tongued
and morticed together, and as there are many miles of
them the total cost up to the present has been over
1,000,000 dollars.

Besides the snow-sheds ponderous wooden bastions
called " cribs " or " glances," have been constructed up
the mountain side to deflect the avalanches, and turn
them into the direction in which they can do the least

possible harm. During our stay in the mountains these were being extended, and new ones erected in all directions.

The snow difficulties are much greater in the Selkirks than in the Rockies; in the latter range the engineers have been able to manage without shedding, but here, owing to the heavier snowfall and also to the difference in the structure of the mountain pass through which the railway makes its way, this great additional outlay has been necessary.

We halted in the centre of a grand amphitheatre of rocky aiguilles and glaciers at Rogers pass. This is not only a "station," which in most cases means little else than a watering tank for the engine, but has a considerable number of wooden houses grouped together and inhabited for the most part by Chinamen, also by section men connected with the railway; by *employés* of the contractors for the snow-sheds, *et hoc genus omne*. Like all these embryo back-of-the-world villages the general observer would fail to find much that was beautiful, either in art or in morals, and the tendency of the inhabitants is to quote Scripture profusely in a sense totally different from what one attaches to it in church. We were glad that we had not to make this place our head-quarters. Four miles further we had to travel, the scenery becoming grander at every curve. Then the train halted at the "Glacier House," our cumbrous luggage was tumbled out on the platform, and our long journey by rail was ended. It seemed a very long day since we left Calgary, though it was now but 3 P.M. In that day of fourteen hours we had left the prairie, ascended and descended the Rocky Mountains, crossed the noble Columbia, ascended to and passed over the summit of the Selkirks, and now were, in company with all the other passengers, quite ready for dinner. When this was disposed of, and the train, with our late companions, gone on its way down the valley of the Illecellewaet,[1] we had plenty of time to take a look round, and as Glacier House was to be our head-quarters for the next six weeks, we must try and give some idea of it and its surroundings.

Glacier House, built on exactly the same plan as the little inns at Field in the Rockies, and at North Bend on the Frazer, is somewhat in the Swiss chalet style and possesses, besides the large *salle à manger* where dinner is served to the passengers of the Atlantic and Pacific express trains which meet here every day, six or seven small but snug bedrooms. One of these we took for ourselves. Another was soon afterwards occupied by Mr. Bell-Smith, a well-known Canadian artist, and the other rooms were occasionally, during our time, filled by guests who stopped off the train for the night and resumed their journey next day. Sometimes they stayed longer, and some most interesting people were amongst those we met. Some weeks no one came, and then again the

---

[1] Pronounced " Illy-silly-wat," meaning " Rushing Water."

little inn was overflowing. On one occasion when we returned from an absence of several days in the mountains, we found that besides our room being occupied, our two spare tents had also been pitched to give sufficient accommodation. After that a sleeping-car was brought up and left permanently on a siding, to accommodate the occasional overflow from the house.

The hospitable manager, Mr. Perley, his wife, and their little niece Alice, about nine years of age; Mr. Hume, the secretary, the French cook and his assistant Chinaman, three capital waitresses and the "boy," made up the staff. Another man, "Charlie," belonged specially to the railway, his chief business being to watch the white stones round the fountains which played in front of the verandah.

These white stones were nothing more than pieces of common vein quartz, broken up to trim the edges of the little ponds; but so impressed were the numerous emigrants who went westward, with the idea that quartz meant gold, that whenever the trains stopped men, women, and children, pounced on these white stones, and would have left not one but for the vigilance of "Charlie." It was a perfect farce sometimes to watch a man sneaking round with his eye on the custodian, trying to steal one small piece, and the very fact of their being so guarded served to enhance their value. The live stock on the premises consisted of a black bear cub, which at first made night horrible by squealing for its mother, but nevertheless was a most intelligent, playful and amusing little animal. Little Alice and the bear were great friends, and until it got too heavy she used to carry it about in her arms. Then there was her cat, some fowl and "Jeff," a most friendly dog, always ready to join in any expedition. Among the various traits and vestiges of ancestry which he exhibited that of Gordon setter predominated. So much for Glacier House. It would be impossible to give any adequate idea of the magnificence of its surroundings; we must however make the attempt.

On leaving Rogers pass the railway commences to descend the valley of the Illecellewaet, but as the river plunges down with a rapidity quite impossible for the railway to follow, the latter has to seek a gentler gradient by availing itself of the lateral valleys, to make wide loops and curves. Glacier House is situated just at the sharpest curve of one of these loops, where the line runs up nearly to the base of a great glacier, and sweeping round, crosses the glacier torrent on a trestle-bridge, and returns to the main valley. That portion of the line opposite the inn is provided with snow-sheds for winter travelling; but to enable passengers to see the scenery, a summer track has been constructed close outside the sheds. The valley below the track, and for a considerable distance, sometimes as much as 2,000

F

feet above it, is clothed in primeval forest. Huge
cedars, pines, and firs draped with curtains of grey
lichen makes up the great mass of the forest, and
shelters almost impenetrable undergrowth, travelling
through which is rendered all the more difficult, owing
to the innumerable fallen trees in every stage of decay.

Down through the forest above the track, a cascade
tumbles in a series of leaps of silvery foam for over
1,000 feet. A pipe from this cascade fills our
fountain and supplies the bath-room, which proved
no small luxury after days of hard travel. Above
the forest-clad slopes, precipices almost we might
call them, and rising above the top of the waterfall,
the crags of Eagle peak stand up 5,000 feet from
the track. Running our eye along this high rocky
rampart, towards the right, that is towards the south,
we behold the huge obelisk of Sir Donald 10,629
feet high, the face towards us a sheet of bare rock
inaccessible to man or beast. Sharp jagged *arêtes* meet
it on either hand and suggest possibilities of ascent.
Farther to the right the lower part of the *arête*
loses itself in a saddle of snow-covered glacier
which forms the sky-line for about a mile. From this
saddle, which we have called the Great Illecellewaet
Nevé, descends in a single plunge of over 2,500 feet a
fine glacier of pure white ice.

The upper portion of the fall is split up into
innumerable *séracs*, the crevasses showing like lines

of blue and delicate green in a surface of silvery white. No moraine pollutes its surface, and it reminded us of the Rhone glacier in its general character. Further to the right, a ridge, clothed with forest nearly to its summit, separates the valley filled with this fine glacier from another which extends for four miles into the mountains, and then all view from Glacier House is stopped by the forest rising immediately at the rear. We must now turn to the left, and there looking out into the main valley and beyond it, Mount Cheops, the Grizzly range, and Mount Hermit rise above the forest in purple crags, bearing great bosses of glacier ice.

Dark green forest, rushing streams, purple peaks, silvery ice, a cloudless sky, and a most transparent atmosphere, all combine to form a perfect Alpine paradise.

Having stowed our heavy luggage in a room set apart for that purpose, there were yet several hours of a fine summer evening to be disposed of, so we started for a walk to the glacier in company with three gentlemen who were staying at the inn over night.

A good path, like many a one in Switzerland, had been cut through the forest. Ferns, yellow lilies,[1] and many other bright flowers enlivened the margin of the glacier stream, as it foamed in wild music over its rough boulder bed. First we crossed, by a firm rustic bridge,

---

[1] *Erythronium minor* (Macoun).

the stream coming from the long valley headed by
the Asulkan glacier. From this to the next bridge,
which spanned the river from the larger glacier, the path
lay amidst the huge boulders of an ancient moraine,
left by the glaciers from the two forks of the valley,
which here met in the days gone by. The forest which
until quite lately had covered the moraine, we found
utterly demolished by a recent avalanche, which had
evidently fallen from the direction of Mount Sir
Donald. The hemlock, balsam, and Douglas firs, though
as stout as ship's masts, had been snapped off close to
their roots; some were torn up and driven long distances from where they grew, and lay in heaps, but
the general position of the trunks pointed distinctly to
the direction from which the destroying avalanche had
come. Even the boulders of the moraine showed signs
of having been shifted, some of them were huge blocks
of quartzite, one I measured was $50 \times 33 \times 24$ feet.
No better illustration could be presented of the over-
whelming power of an avalanche, though composed of
nothing else than the accumulation of a winter's snow.

Crossing the second bridge the path made a few zigzags up through the forest, and then skirted the hill-
side through meadows of coarse grasses and the large
succulent-leaved *Veratrum viride*, a most striking and
characteristic plant in all these valleys, wherever the
heavy forest or the alder scrub fails to establish itself.

From this place the path forks, one track leads across

"The forest we found utterly demolished by a recent avalanche."—P. 68.

the boulders to the foot of the glacier, crossing innumerable streams of milky water. The other con-
tinues up through alder scrub to the higher slopes,
commanding fine views over the ice with its blue
fissures and crevasses.

H. and I chose the upward path. Mounting rapidly,
and crossing a stream coming down in a cascade from
the glaciers to the left, we reached the farthest limit
of the path, and sat down to study the scene. The
pine forest, though extending for over 1,000 feet on
the mountain sides, above the lower portion of the
glacier, ceased in front of its terminal face at a distance of about half a mile. This was evidence of the
retreat and shrinking of the glacier.

The pines, where the forest ceased abruptly, were at
least twenty years old, the space between them and
the ice being covered by alders. The advance of the
forest in the track of the glacier does not seem to be in
direct proportion to the rate of its retreat and the age
of the trees. The rate at which the glacier bed can
crumble itself, and be prepared by the leaf mould
formed by the alders for the growth of pines and firs,
is a process so slow that pines have time to grow to full
age, and fall, and rot, before the soil ahead is prepared
for their advance. Not only here, but in many other
valleys we were in, the same thing was noticeable. The
alders thrive where there seems to be little or no soil,
and they can also, owing to their pliancy, hold their

own in avalanche tracks, where the stout pines would be utterly smashed up. However useful the alders may be in performing these offices in the economy of nature, they form an almost impenetrable jungle, and one of the greatest obstacles to travel that it is possible to imagine. On the steepest slopes they grow downward, and after a few feet turn upwards to the light. Scrambling through their stems thus involves stepping over a selection of stiff springing branches, and stooping sufficiently low to get yourself and your pack under the next branch above. You can, when in such scrub, but seldom get your feet on the solid ground, so a slide downwards, and a sudden wrench on your arms, in trying to check your descent, is a matter of constant recurrence.

As the shades of evening were closing in we thought it wise to return to our quarters. Having wandered a little beyond the path we missed it for some distance, and got our first experience of trying to take a " bee-line " through the alders. The first blood was drawn, the first skin knocked off our shins, before we reached our supper and bed. Yet we went to sleep with the satisfaction that no time had been wasted; for we had taken our first walk in the Selkirks on the 19th day after leaving Queenstown.

CHAPTER VI.

" On the over-worked soil
Of this planet, enjoyment is sharpened by toil ;
And one seems by the pain of ascending a height
To have conquered a claim to the wonderful sight."

OWEN MEREDITH.

Our map.—First climbs.—Packing.—View from *arête* of Mount Sir
Donald.

OF the peaks encircling and within sight of Glacier
House, some had names. These soon became familiar
to us, but of what lay beyond the ridges forming
the sky-line, no one could give us the faintest
idea. Mr. Hume, the assistant manager of Glacier
House, had accompanied some gentlemen during the
winter, on a snow-shoe expedition to the head of the
Asulkan valley, and obtained a view of the ranges
beyond, but his ideas of the topography were not
very clear. Few people, accustomed to visit Switzerland, where accurate maps are found ready to hand,
by which they can unravel the maze of ranges and
valleys seen from a mountain top, can have any con-

ception of the difficulties of forming a definite picture
of a rugged country which has never been mapped
out, where one's outlook is closed in by higher mountains, and where but little disconnected bits of valley
are visible. The riddle we had undertaken to solve was
the structure of that section of country lying immediately to the south of the Canadian Pacific railway
track and inclosed by the highest peaks of the Selkirks.
During the first couple of weeks' work, nothing seemed
to develop clearly.   Then the scene began to take
shape, and when the time came for us to leave the
mountains, we found it hard to recall those first feelings
of bewilderment, so familiar had the mountains,
glaciers and valleys become.

In undertaking any topographical survey, the first thing
to decide is,—On what scale shall we make our map?
For many reasons we came to the conclusion that four
inches to the mile would be sufficiently large on which
to put down all the details of importance. The next
step is to measure a line which shall form a base for our
first triangle. If a base-line of one mile is measured,
the picture of that on the map is a line of four
inches, and thus the scale of the map is fixed; all other
distances being in equal proportion.

In great trigonometrical surveys, such as that of
Great Britain and Ireland; the measurement of the base-
line is of such importance that it is done with micro-
scopic accuracy.   In our case such minute accuracy

would have been impossible, so we were content with a measurement made with our steel wire one-eighth of a mile long.

One engineer said to me, "But surely with such a measurement you must allow for the expansion and contraction of the wire with varying temperature." All I can say is that I made no such allowance. The temperature of the day when I measured the wire and of those on which I used it, did not differ much over 10°, certainly not 20°, and if there is no greater inaccuracy in our location of peaks than the error arising from the expansion and contraction of our wire I am satisfied. A much more fruitful source of error arises from the difficulty of always being certain that you are observing the same particular knob, on a mountain summit, when seen from different points of view. This is particularly the case when the peak occupies a plane much above the point of observation. In a mountain survey therefore the points fixed from the highest elevations are the most reliable.

When breakfast was disposed of on July 18th, we packed the plane table on our shoulders and set off for the opposite side of the valley ; the top of the snow sheds there, being the only level place where it seemed possible to measure a straight line of over a hundred yards in length. There was even here a slight gradient (which was unfortunate), but as no better spot could be found, we measured 660 yards and set up a

pole, with flags at either end. The flag at No. 2 station
was a piece of newspaper, and as it fluttered from its
staff for over a week, some idea may be formed of
the great calmness which prevails in these mountain
valleys. From points at either end of our base-line we
fixed a third station, at the opposite side of the loop
made by the railway, and took, on the plane table,
bearings of all the peaks in view, made profile sketches
and photographed them, assigning them numbers for
future identification. To accomplish all this took the
greater part of the day, and as the sun was shining
with intense heat and mosquitoes were biting like
fury, we were not sorry when it was completed.

The cascade which forms such a feature in the view
from Glacier House is no small source of difficulty to
the railway people, as it objects to be controlled in
any way. A bridge has been built for it to
go under, but with the true spirit of freedom it uses
the bridge only occasionally, and just then was with
much hilarity dashing right down on the railway and
knocking away all foundation from the track. By
balancing ourselves carefully on the rails we crossed it
without much difficulty; and H. lay down with his
chest on the rail to regale himself with a drink. Suddenly, to my horror, a freight train, coming down the
gradient, swept round the curve. Men stood on the
roof of every car screwing down the brakes. The
whistle was evidently blowing, but the cascade drowned

all other sound. I shouted to H. "Here's the train."
He took no notice! There was no time to speak
twice; but at the very last instant he perceived
the danger and rolled himself aside just as the
train roared past. After this we were careful not to
lie down with our heads on the main track again.

Having completed our preliminary observations, the
next object was to reach some of the higher points
we had observed, and so extend our survey. The peak
separating the great glacier from the Asulkan valley
seemed the most central, so next morning, July 19th,
we started for our first climb, carrying the plane
table and a half plate camera.

A short distance from the inn, and just beyond
where the forest had been demolished by the great
avalanche, we left the path and struck straight
up the mountain side through the heavy timber.
After the inaction of the voyage, and the long railway journey, we were of course in no kind of training,
so the steep ascent had to be taken slowly; and as
our shoulders ached under the packs, we were compelled to rest pretty often; in spite of the mosquitoes,
who seized on these moments for most furious
onslaughts.

Fallen timber lay in every stage of decay, covered
partially with rhododendron and blueberry bushes.
The whole ascent was one continuous scramble,
the bushes giving us the means of hauling ourselves

upward. For 2,000 feet we ascended through forests so dense that no distant view was possible. Nothing was visible but the huge stems of the hemlock and balsam firs. At an elevation of 2,000 feet above the railway the trees become more gnarled and dwarfish ; in shady hollows we came on patches of snow, the melting of which furnished us with a drink, for which we were famishing after three hours of climbing, under the close heat of the forest.

Next we came to some grass slopes, and after a scramble over great heaps of shattered rocks, composed chiefly of a conglomerate which cropped out at a gentle angle dipping to the south-west, we reached the knoll forming the apex of a triangle which we had plotted on the previous evening. Though not actually the summit of the ridge, it was a good clearly-marked position, commanding a splendid view of Sir Donald and all the surrounding mountains, and according to hypsometrical and barometrical measurement was 3,500 ft. above the level of the railway line at Glacier House. We remained here for two and a half hours, taking observations, photographing, collecting specimens of *antennaria* (something like edelweiss) and other flowers, which were just showing their heads amongst the rocks where the snow was melting. Looking eastward the great fall of the glacier formed a fine foreground, and beyond it Sir Donald rose in bare precipices.

We scanned these walls of rock most carefully, to see
what chance there might be of ascending the peak, but
decided that, from our present point of view, no possible
route was apparent. We were, however, able to plan
an ascent of the further margin of the glacier, to the
great swell of *nevé* which rose higher than our present
level, and by ascending which we might get round to
the south of Sir Donald, and possibly find some practicable route in that direction.

Having made what notes we could and enjoyed a
respite from the mosquitoes, we commenced our descent.
A short scramble over rocks and a glissade on a
snow slope, brought us to the margin of the forest,
through which we plunged and slid, and scrambled,
knocking a good deal more skin off our shins and tear-
ing sundry reefs in our garments. It was not easy to
keep to the track by which we had ascended, consequently we soon found ourselves amongst precipitous
rocks clad in moss and fern, and as rich in London
pride and other saxifrages as some well-known
spots in the hills of Kerry. We had to skirt these
precipices and find some way or other down through
them. They were not high, for the tops of the
pines growing at their foot mingled with the lower
branches of those growing above. The jolting of the
packs of instruments on our shoulders, unused for long-
time to such work, was tiresome in the extreme, but
at last we had descended 3,000 feet and were in the

more level forest.   Selecting the best line we could, we
made our way out on to the open boulders of the
ancient moraine, and taking a course through the *débris*
of the great avalanche, we reached the path, and in a few
minutes were enjoying a refreshing warm bath at our
hotel.

Our ascent proved most helpful in making future
plans, and as I had now material enough to commence
on, I devoted all the following forenoon to plotting
out our observations.   H. meanwhile went off down
into the valley below the railway to hunt up the route
he had followed in 1884 and take a few photographs.

Before we could attempt any distant expedition,
help in the way of porters of some sort was essential.
We had asked Mr. Marpole, the superintendent at
Donald, to try and get us a pack-horse. Late on
Sunday night we heard the whistle of a locomotive,
and next morning found that it had brought us a
good-looking cayeuse, or Indian pony, which promised
to be a useful helper, long scars on its flanks
showing that it had already seen service in the
mountains. Ever since our arrival we were engaged
in correspondence with people most likely to find
packers for us.   Our letters and telegrams proved
abortive—one very good man we had written to
would have come but that he had just taken another
job, he lived on the eastern slope of the Rockies.
Then at Donald some mighty hunter was discovered

who expressed a desire to join us and accepted our
terms, but when he heard we were two parsons he
"chucked it up" in disgust, saying that he would
have to knock off swearing for over a month and that
that was utterly impossible. A man while we were
at Glacier House undertook to guide an American
party up the glacier. One of the gentlemen pointed
out to me the route they proposed to follow, it was
beset with great dangers which none of them seemed
to realise. Their guide carried an American axe over
his shoulder and their lunch under his arm. I asked
him what he wanted with the axe, he said "To cut
steps!"

Not wishing to pose as general adviser I said nothing,
but offered my ice-axe to the elder gentleman, the
younger having decided on taking his gun. So off
they set. A few hours later they came back
thoroughly scared. The "guide" had slipped and was
within a hairsbreadth of losing his life. The axe
and the lunch vanished into a crevasse, never more to
be seen, and the gentleman who had my axe slid
seventy feet, and would have been into a crevasse
but for the axe, which fortunately for me and for
himself he held on to like a man. That "guide" was
a very good fellow in many ways, but after this
adventure he considered glaciers were not in his line.

Now that we had the horse we determined to make
a move, and our first thoughts lay in the direction

of Mount Sir Donald. To explore the side we had
not yet seen, that facing the south-east, was our object.

On Monday July 23rd we selected camp material
and provisions, and invited Mr. Hume, the secretary,
to join us in our climb. When the trains had come
and gone and the mid-day meal was over we were
ready to start. Charlie too was able to abandon his
care of the white stones, and came to lead the horse
back when we reached the furthest point possible
for him to go.

The cayeuse was now sent for to his stable and
we commenced to pack. No easy matter was this
packing, for no pack-saddle had come with the beast,
so we had to do our best with a riding saddle. We
adopted a plan that I had used most successfully
in New Zealand; laying a canvas hammock across
the saddle, we placed a pack on each side, turned
up and fixed the ends of the hammock at the top,
and this, with a synch or belly-band round it, does fairly
well on a pinch.

The horse had however his own notions as to
packing; he turned round the whites of his eyes, most
suspiciously laid down his ears, and the moment the
pack touched him, he backed, kicked, and lashed out
in an alarming manner. After much coaxing on
our part and kicking and biting on his, we fitted
the packs on and started up the valley, in Indian
file.   H. and I carried our rifles, and on our backs,

knapsacks containing the surveying instruments,
cameras, &c. All went well for a mile, and as the
horse was going along as quietly as a cow and the
evening was warm, I saw no harm in hitching my
knapsack on the pack. H. also relieved himself of
his heavy rifle, and we walked on happily, through the
meadows of *veratrum*, up by the east bank of the
glacier. The path was getting steep and the pack
seemed to need bracing up. Some idea of a similar
nature must have crossed the mind of the cayeuse, for
without the slightest warning, he took a sudden fit of
buck-jumping, tumbled down, rolled over and over
down the slope, and when our goods were thoroughly
mashed up and scattered to the winds, he got on his
legs and shook himself with apparent satisfaction. It
was really too horrible—I rushed to my unfortunate
knapsack. If the man from Donald had been with
us I think I'd have given him permission to swear
for five minutes without stopping, and so vicariously
relieve my over-burthened mind. A sextant, fortunately not a new one, was smashed to bits. I picked
up its little ivory scale all by itself on a bush. A
thermometer which had been carefully tested at Kew
was in shivers. I could not look at my photographic
plates then, and concluded they were all broken.
Fortunately however they escaped; the rifle too came
off all right. But oh! what fools we felt at having
been taken in by the deceitful calm of that cayeuse's

temper. We all made good resolutions on the spot, and
kept them so far as never again to trust any instrument
to the tender mercy of a horse.

After this piece of diversion was over, the cayeuse
seemed quite satisfied, and allowed us to pack the load
again and proceed on our way. On reaching a stream,
descending from the high glacier on our left, we considered that the horse could go no further, so taking the
packs on our own shoulders, we sent Charlie home with
the nag. "Jeff" the dog had come with us so far, and
now refused to return. We did not want his company at
all, but there was no getting rid of him. Fording the
cascade we ascended into some forest, beyond which a
fine waterfall, formed by a stream breaking out from
the side of the great glacier, fell vertically about one
hundred feet over a rocky cliff. Not having explored
the way ahead, we lost some time seeking out a site
for our tent, but after an hour's scrambling over the
boulders near the fall, we found a level hollow in the
forest not far from the top of the waterfall, where
throwing down our packs we lighted a fire, pitched the
tent, and made ourselves snug for the night. While
cooking supper the smoke kept the mosquitoes away
and the netting kept the tent safe from them, but
our necks and hands were in hills and hollows from
bites. H. said that he had a complete model of the
Rocky Mountains round his neck.

Owing to our late start it was quite dark before

supper was finished and we were ensconced in our
sleeping-bags, trying to sleep. Poor "Jeff" seemed to
be much troubled by mosquitoes, we could hear him
snapping at them, and his low whining howl now and
then suggested either bears or mosquitoes, we knew
not which.

We got up before dawn, lighted the fire, fried the
bacon, made our tea ; and when breakfast was over we
carefully extinguished the fire, and at 4 A.M. started
upwards. A few yards from our camp we emerged
from the forest and commenced a long scramble over a
waste of huge boulders.

Though the opposite side of the valley and the
upper swells of the glacier were touched by the golden
beams of the sunrise, we still enjoyed the cool shade
of night.

The valley far below was hidden in filmy white
mist, the chatter of the streams and subdued boom of
the waterfall made sweet mountain music.   Occasionally the harmony was broken by the loud crash of
a falling *sérac* on the glacier, or the wild, shrill whistle
of a hoary marmot. A cry so wild and strange that
we were startled by it in our earlier climbs, but we
soon got used to it, and now I can never think of one
of these wild valleys without the cry of this large
marmot sounding in my ears. There were a number
of other animals living in these boulder heaps, and
"Jeff" was death on them all, at least so far as his

hopes and imagination went. He was off, charging
over the boulders the moment a whistle sounded, but
as the animals generally gave their cry of alarm at the
mouths of their burrows, they were pretty safe from
" Jeff," and he most successfully prevented our having
any chance of a shot at them. We wished " Jeff" in
the happy hunting grounds, where, no doubt, marmots
have no burrows to retreat into. After an hour's ascent
over these great boulder heaps, we reached the first
slope of snow, and as there was no likelihood of any
game higher up, H. cached his rifle. Snow slopes
and boulders alternated for another hour, then came
a steep snow ascent, and at 8 A.M. we reached Perley
rock. This rock formed an islet in the ice-world
close to shore, and was covered with sparse vegetation ;
some pretty heaths suggested firing should such
be required. Above this we were fairly on the
glacier, and as crevasses were visible we put on the
rope. The view ahead was confined by the rising
swells of ice. The view backwards and away over
endless ranges to the north and west was wonderfully
beautiful.

We had now turned the south-western shoulder of
Sir Donald, and while ascending some very steep slopes
where we had to cut steps in the hard frozen surface,
we were able to turn away to the eastward, thus getting
towards his southern face. The slopes now became
more gentle, so we sat on our axe-heads for a few

minutes to rest, and "Jeff" sat on his tail, nothing
daunted. Sometimes he followed in the steps, and
then he would try a rush up the slope, and with much
scratching and many slides downwards, return once
more to the slow and sure method of ascent. We had
now got into a regular bay in the back of Sir Donald ;
a rock *arête* ran down to the south-east, and another
to the south-west. The summit of the main peak
looked quite close over the apex where these *arêtes*
met, so we faced up for the tongue of glacier which
filled the bay, and sloped up to the crest of the north-
west *arête*. There was of course a bergschrund across
the foot of this ascent, and another big crevasse a
little way up. Before tackling this slope, which promised
some tough work, we halted for lunch, and cached a
small tin of meat. The bergschrund gave us no trouble
as it was well bridged, so we worked up for the
crevasse. On reaching its lower lip we found we had
to follow it along towards a bridge to our left, and this
brought us right over the most open portion of the
schrund below, so we had to go very cautiously. The
slope was steep, and every bit of snow we dislodged
went singing away into the abyss below. The snow
was soft and we were able to kick well into it, but
in two places it had recently slid, which was another
cause for anxiety. As for "Jeff" the grip was just
right for him. I only envied him in his blissful unconsciousness of schrunds and crevasses. He was wiser

however in these matters than we could have imagined, as we learned later on.

A few hundred feet of slope, and the crest was reached. Here the snow curled over into a cornice and overhung the great bare smooth precipices, down which our view extended to the lower glaciers in the hollows of Sir Donald, and still farther down to the dark pine forest, and the railway, like a tiny hair-line, winding about through the valley 5,000 feet below us. As yet our prospects of reaching the main peak were uncertain, so we tackled the snow *arête* leading to crags where this south-west ridge met that running down to the south-east. The view beyond these ridges was completely unknown. Anxiety as to what that unknown side of the mountain was like hastened our steps upward.

I had to cut a few steps, for the icy cold wind on the ridge kept the snow frozen. Then the axe clinked on the lichen-covered quartzite crags, and in an instant the anxiously expected panorama opened before us.

We were on a pinnacle of rock, according to the barometer just 10,000 feet above the sea. On all sides were vast precipices, and down these precipices our eyes ranged to the green, forest-clad valley of Beaver Creek, the river being visible for many miles, winding with an infinity of curves 6,000 feet below us.

Beyond the river rose a range of hills with flattish

plateaux on the top, flecked with snow. Still further
to the eastward, range rose upon range, fading into
purple and blue. Above them all, the Rockies, bearing silvery white glaciers, formed a sharply defined
sky-line, and were visible for over 150 miles. This
wonderful panorama constituted our view to the eastward. To the southward it was totally different; in that
direction the undulating fields of glacier lay like a
great soft white blanket, covering up everything for
ten miles, beyond which other snow-seamed crags rose,
rivalling, probably in some cases surpassing, Sir Donald
in elevation. To the westward other ranges were to
be seen, and one huge ridge of black precipices capped
by ice rose high above the glacier and seemed to
dominate the scene. Its foot was separated from us by
two intervening ranges, and appeared so difficult of
access that we felt but a very faint hope of some
day reaching it. This mountain we named after
Professor Bonney, and we afterwards scaled it successfully.

Beyond the valley of the Illecellewaet to the north-
west, some fine peaks were visible; one dark, bare
rock pinnacle bearing north-west, was most striking,
and no doubt over 10,000 feet high. Our view to the
northward was blocked by the last great crag of Sir
Donald, from which we were cut off by a notch
200 feet deep. At its bottom a narrow rock *aréte* joined
the precipice below us with the face of the final peak.

Below this *arête* on one side lay the glacier visible from Glacier House, and on the eastern side in a deep hollow, a fine glacier which we named the Sir Donald glacier, commenced its course, and flowed outwards in beautiful fan-like structure, in the direction of Beaver Creek. The cliffs rising at the farther side of this latter glacier, that is in the great buttress supporting Sir Donald from the valley of the Beaver, were so steep that not a speck of snow clung to them. Had we had a rope about 200 feet long, we could have descended from our perch and then easily crossed by the little connecting wall to the main peak, but the face of it looked about as inaccessible a piece of rock as any climber could wish to see. There were a few cracks and ledges which may one day be used by some one, but " the quest was not for me."

During our ascent we had been more troubled by heat than by cold ; now however a strong icy wind blew from the north-west. H.'s hat, or rather one of mine that he had affected, fluttered off cheerfully to Beaver Creek, and when we came to the plane table-work I had to slap my arms to get life into my fingers. As the summit we were on is a little peak plainly visible from Glacier House, we fixed a red handkerchief to a crag, and having secured nine photographs, began the descent.

In taking the first snow steps we had to go backwards carefully, as it would not do to slip ; the

background to our feet, as we cautiously watched
them safe into the steps, being the valley of the
Illecellewaet, 6,000 feet below. As I was the last in
the descent I buried my axe deep, and watched to
keep the rope tight. H. cleared the steps below. Mr.
Hume was in the middle, and "Jeff" kept just above
me, with his nose held low and his legs spread out
wide. He wished, I think, to get a grip with his tail
too, but as that was impossible, he satisfied himself
with the fact that he could not slide farther than
against my legs. On returning to the place where we
first struck the *arête* we continued the descent, still
keeping our faces to the snow and getting a firm grip
with the axes at every step. We took these precautions on account of the crevasses below us, otherwise
we might have made a splendid glissade. Now that
we were away from the brink of the great precipice
"Jeff" got tired of his precautions, and considered a
charge down the slope would be pleasanter. It was
most ludicrous to see him trying to stop himself. Legs
and tail in full action, and he all the while swinging
round and round on his vertical axis. He was determined however that no horizontal rotation should
follow; if it had, the days of "Jeff's" rambles would
have terminated there and then. Having crossed the
bergschrund we glissaded in safety, and regained some
provisions we had cached. On reaching the flat surface
of the glacier I measured a base-line of 300 yards, for

the purpose of fixing the location of some points on the glacier. Then we resumed our descent. Several gl,ssades brought us down to the boulders at the side of the great glacier fall, and passing our camp without stopping, we reached Glacier House at 5 P.M. after thirteen hours' work.

CHAPTER VII.

"The mountain-ranges are beneath your feet. . . No trace of man now visible ; unless indeed it were he who fashioned that little visible link of highway, here, as would seem, scaling the inaccessible, to unite Province with Province."—CARLYLE.

The railway gangs.—The pack-horse again.—Sledging.—The valley beyond the snowfield.—Camped on Perley Rock.

THESE mountain railways give employment to a great number of men. About every five miles of the track is under the charge of a special section gang of ten or twelve navvies and their "boss." A snug log-house is built for their accommodation, and the wives of the married men look after the cooking and washing. The men go to their work, when at all distant, on a truck propelled by a mechanical arrangement worked by pump handles, and the overseers have tricycles which also fit on the rails; two wheels are on one iron and a third small wheel at the end of a slende outrigger rests on the other; the rider sits over the pair of wheels on one rail, and propels himself at

great speed by handles which are worked like rowing
a boat.   The station agent at "Glacier" had one of
these. We used sometimes to take a spin along the
track on it, but it came to an untimely end, fortunately
when in the owner's charge. He was going up through
one of the snow-sheds when he met the train coming
down the incline. He had just time to save himself
by leaping off, but the tricycle was knocked to bits
by the cow-catcher of the locomotive. Besides the
section gangs, each trestle-bridge has its special watcher
who gets three dollars per day, lives in a little hut
by himself, and whose business it is to examine the
whole length of the bridge after the passage of each
train, to see that no injury has occurred or that it has
not caught fire. Along the bridge, on top, are a row
of barrels of water which he can use for extinguishing
fire. Should however anything serious occur he has a
telephone in his hut, by which he may communicate
with the section gangs on either side of him. The
snow-sheds are also carefully inspected after the
passage of every train, and they are usually supplied
with a complete system of water-pipes and coils of hose
in case of fire. Besides these gangs of men specially
connected with the railway company, the contractors
for building snow-sheds, cribs, and bastions, or glances
as they are called, for protection against avalanches,
had their gangs. Some of these men were employed in
the dangerous occupation of felling timber on the

steep mountain sides and shooting the logs downwards.
One fine fellow was killed at this work and several were
injured while we were in the Selkirks. These men get
three or more dollars per day, and their board costs
less than a dollar per day, so for steady men there is
here a good chance of saving money. These gangs were
composed of men of all nationalities, Italians and Swedes
seemed however to predominate. Casuals occasionally
passed along the track by way of seeking work. They
were usually of rascally appearance, and though we
were hard up for packers, these unemployed were the
last to look to for help.

Since our expedition up Sir Donald I had plotted out
our observations; and the conclusion I came to was,
that our next exploration must be across the great
glacier field, in order to see what the ranges and valleys
beyond it were like. This would necessitate our carrying a camp outfit and provisions over the summit ridge,
and this without further help would be impossible.
The superintendent at Donald kindly gave me permission to take men from the section gang for this
purpose, if any were willing to come. Two respectable
young men volunteered from the gang near "Glacier,"
and on July 26th we packed our larger Alpine tent
with sleeping bags for the four of us and provisions for
about a week, into as small a compass as possible, and
by mid-day we were all ready to start. The weather
during the morning looked threatening, and at noon

thin films of mist drifted across the sky. The peaks looked angry with torn clouds, and a flash of bright lightning, followed by a deafening crash of thunder, formed the prelude to a fierce thunderstorm accompanied by a deluge of rain. It lasted for only an hour; the sky quickly cleared, the sun shone forth, making every spine of the firs sparkle with rubies and emeralds. At 3 P.M. we sent for the pack-horse, and determined to move on to our tent in the woods, near the foot of Sir Donald, for the night. The cayeuse proved more docile than on the former occasion, and we fixed the packs on him with little difficulty and wended our way towards the glacier. Crossing the rivers we commenced the steeper portion of the ascent. Suddenly, as though some unseen terror haunted that particular spot, at the very same place where the disaster had before occurred, the cayeuse was seized with a paroxysm of buck-jumping; the packs flew off, he rolled down through the ferns and rocks, and then, perfectly satisfied with his performance, stood patiently, while we restored our goods to his back. The instruments were this time safe in our knapsacks on our own backs, so the harm done was *nil.* "Jeff" of course came with us, but as we did not wish for his company across the glacier, on account of his scaring the animals which we hoped to shoot, we insisted on his returning with the cayeuse and Charlie from the cascade, where we took the packs on our shoulders.

Now that the way was familiar to us, we carried our
loads upwards with less difficulty than formerly, but
as the amount of blankets, provisions, ammunition, &c.,
weighed a good deal, and as our two companions were
not capable of carrying heavy loads, like Swiss guides,
we had to divide our burthens into two packs each
and return over the ground for our second pack when
we had carried the first on a few hundred yards. Thus
we gained our camp in the forest, and found everything
in statu quo.

It was not yet dark, so to save time in the morning
we carried upwards a pack each, and deposited these
under boulders beyond the forest line. Then we returned
to camp and cooked our supper. As all the bushes
were dripping wet after the thunderstorm, we had
to split some logs to get at the dry inside wood ;
soon the fire crackled merrily and the smoke put the
mosquitoes to flight. This tent (the tent left standing here since our last expedition) being all ready for
occupation, we did not open our packs to get at the
larger one, but as it was only made to hold three, a
party of four filled it almost to bursting. We lay like
herrings in a barrel, head and tail, and needed little
covering for heat.

In the morning we found it necessary to rearrange
our packs, so it was 6 A.M. before we extinguished the
fire and proceeded on our upward way. We were not
long in reaching our cache beneath the boulders, but

to our great disgust found that some beast had got at
our packs, and with a most depraved taste had breakfasted off my Alpine rope. It was on the outside of
a pack, tied on in a coil, and the wretched creature
had nibbled through every bight of the coil, thus
cutting up the rope into a number of short lengths.

Throwing away the shorter lengths which were no
good, and consoling myself with a feeling of extra
generosity in thus providing the beast with a dinner,
I put up the best pieces carefully, and at our next
halt for rest, spliced them together. Our course now
lay up the huge piles of boulders at the side of
the glacier. These were composed of hard quartzite
and micaceous schists, some of the latter shining with
a beautiful silky lustre, and suggesting by their lamin-
ated structure, the great squeezing they were subjected
to in the thrusting up of the mountain masses. As
these schists were more friable than the quartzite, the
configuration of the district seemed to depend to a
great extent on the disintegration and denudation of
these softer schists and the permanence of the harder
quartzites in the mountain ridges. Between the boulder
heaps were patches of heaths and grass all gay with sub-
Alpine flowers; yellow lilies, the bright scarlet flowers
of *Castileia miniata*, the large purple daisy *Erigeron
macrantha* and *Anemone occidentalis* (like *A. verna* of
Switzerland) were amongst the most striking. Ferns
lurked in the crevices between the boulders; there we

found amongst others the holly fern, *Aspidium Lonchitis.*

Now that " Jeff" was absent, we saw more of the " whistlers," as the large hoary marmot is frequently called; the little chief hares, about the size of a rat, were common, and we succeeded in killing one with the spike of an ice-axe. A beast called the Sewelell also lived in these boulder heaps, and is remarkable for its habit of collecting flowers. While we were wending our way upwards H. holding up a bouquet of flowers sang out to me, " Some one has been up this way before us." I knew that this was not the case, but was fairly puzzled by meeting another bouquet of plucked flowers with the stems laid neatly together as though some child had laid them down. Afterwards the mystery was solved by finding bouquets near the burrows of these animals. These flowers, though growing in regions till lately unknown to man, have not been wasting their sweetness on the desert air. Besides the mountain bees who drank pleasure from their cups, these Sewelells seem to have fully appreciated their beauty or their sweetness in some strange way, and collected them while the summer lasted for a winter store.

We saw a few grouse, and while H. went back for his second pack, I took the walking-stick gun which he carried, and determined to secure one of these fine birds for the pot. Having stalked it most carefully, I held

H

out the gun to glance my eye along the barrel, and
when " on " the bird pulled the trigger. I know there
was a loud bang, but what was impressed much more
forcibly on me was a blow on my nose from the recoil,
which brought all the stars of the firmament into my
eyes, with the sun and moon and a few comets into the
bargain.   It was a shame for H. not to have warned
me that the charge he had put in was big enough for a
grizzly bear.   I don't know what became of the grouse.
I never fired that gun before and *never* since.

Going over all these steep boulder heaps twice with
double packs was fatiguing work, and about mid-
day we halted beneath a large block of quartzite,
which promised us some shelter from the sun's rays.
Presently a loud shrill whistle from the other side
of the rock made us aware of the close proximity of a
marmot. We sat quite still, while H. with his walking-
stick gun crept round the corner to reconnoitre. Soon
the hill-side reverberated with a bang, and amidst a
cloud of dust and stones, the marmot tumbled from
right over our heads, and rolled some twenty yards down
the slope. H. soon reappeared ; his nose bleeding and
complaining that he had missed the beast, that it had
run somewhere, where he could not see, and that he was
not to be blamed as he had had his nose smashed.
He was therefore immensely surprised when we
pointed to the dead marmot on the rocks below, shot
through the neck.

As we were to camp on Perley rock, still nearly
1,000 feet above us, and in the midst of snow and ice,
it was necessary to consider the question of fuel.
Though we were far above the forest line, some dwarf
balsam pines managed to sustain life in the face of the
cliffs above the boulder heaps on our left. Taking the
axes and rope we spent an hour scrambling on these
ledges, and pitching down dead twigs to the snow
below. Then we fixed some to each pack, and proceeded to ascend the snow slopes to our eyrie for the
night. Perley rock was separated from the mountain
side by a tongue of glacier covered with firm snow.
Lower down than where we got on to it this became steep
bare ice, and so joined in with the main ice-fall of the
great glacier. The snow was in splendid order, just soft
enough to kick our toes into it, and firm enough to give
good steps. Crossing it as soon as possible we took to
the crags, and finally gained the flat summit of the rock
at 5 P.M. We were not long in pitching the tent, but the
scanty tufts of heath gave us but little bedding, though
we collected every bit we could find. When supper
was over we sat on a prominent crag, and tried to take
in the full beauty of the scene. For over 2,000 feet
below us, the great glacier poured down its grand ice-
fall into the blue darkness of night. The gloomy forests
made the valley look still darker, and we wondered
whether any one at the inn was watching our little
column of smoke, as we sat up aloft in the golden light

of the sunset. The sky was almost perfectly clear. The few clouds which clung to Eagle peak, and other un-named summits, were sharply defined, rounded in outline, tinged on one side with rose colour, and on the other with rich purple shadows. Sir Donald and all the higher peaks glowed for a while in this rose-coloured splendour, and as the sun sank they assumed the silvery grey of night.

We were able from this point of vantage to make out the structure of the valley beneath us. What a wonderful rift it seemed! The cliffs of Eagle peak told their part of the story very distinctly. The strata there dipped in a curve to the eastward. We knew from previous observation that the rocks on the oppo-site side of the valley dipped to the westward, and so the steep scarped faces forming the sides of the valley must be the sides of a huge crack that split an anticlinal, high arched, fold of rocks from top to bottom. Frost and heat, ice and rain, then went to work and pro-duced the valley before us. This structure was only visible so far as this branch of the Illecellewaet valley was concerned.

Immediately after the sun had set, the downward draught towards the valley seemed icy chill, passing as it did over many miles of snow; so we huddled into our tent and tried to forget the hardness of our bed. Our companions did not like it much, and seemed to think the bunks in the section-house were far preferable. However we did not ask them to endure it long, for there was bright moonlight, and not knowing in the least what lay before us, we were anxious to get off as early as possible. The air was too cold to make standing about agreeable, so after a hasty breakfast we packed up the tent, blankets, and provisions.

From the rock on which we camped, a little *arête* of snow led to the first swells of the great snow-field. Along this we carried the packs on our shoulders, and then extemporising a sledge out of the tent poles, across which we lashed the legs of the plane table, we placed the packs on top, and harnessing ourselves by the Alpine rope, which I had spliced together, we set off for the summit ridge. The snow was frozen hard, so the sledge ran easily over the crisp surface. After a little ascent our course led us into a depression where a few crevasses were visible, so H. went behind and with our second rope held the sledge in check. Passing through the hollow we ascended by easy slopes to the summit ridge. The sun was now up, and as there was no cloud in the sky the reflection from the spotless surface made us glad to put on our blue goggles.

At 7 A.M. we were on the summit from which the glacier began to slope in great bosses of ice, cleft in some places by huge crevasses towards the south-west and south-east. The summit ridge was a saddle of pure snow, quite flat for about two miles. And as it commanded views

of well-known peaks as well as of the unknown valleys
and peaks beyond, I measured a new base-line of
660 yards, and took plane table bearings from either
end. While I was thus engaged with our men, H.
boiled the thermometer, and determined afterwards our
elevation to be 8,729 feet above the sea, or about
4,600 feet above the railway at " Glacier."

These observations occupied us for over an hour and a
half, after which, sitting on the sledge, having a feed, we
discussed our next move. To descend by one glacier
or other to a new camping ground would be desirable,
but with so many crevasses in sight, and only the first
portion of the descent visible, we determined that
there was no use in hauling the sledge downwards,
till we had explored the way ahead, and found some
practicable route. Leaving our companions at the
sledge, H. and I threw off our coats, tied ourselves
together to the lightest rope, and taking nothing but
our axes and a prismatic compass, set off down the
glacier towards the south-west. For a mile we went
as fast as our legs could carry us, then crevasses
yawned ahead and we had to go cautiously. We were
already about 500 feet below the sledge. The descent
now grew more rapid, big *séracs* showing beautiful ice
stratification in their sides, were piled in wild confusion. We were driven to make wide zigzags, in
search of safe bridges of snow. The cliffs bounding the glacier on either hand began to assume a

grander aspect. Those on our right were overhung by
a heavy mass of glacier, a prolongation at a high
level of the great snow-field; from which small and
large pieces of ice were continually rattling down,
making it unsafe to follow the margin of the glacier.
Between us and the cliffs of the peak I have called
Mount Fox,[1] on the opposite side, there was a well
nigh impracticable ice-fall. This ice-fall would not
have stopped us, however, if the track beyond looked
promising. As yet we could not see down into the
valley; the ice rounded off so gradually that it always
shut out from us the portion of the glacier immediately
beneath us. The distant portion of the valley was
however quite visible, and looked most interesting.
Deep down in the gloomy shadows of the cliffs of
what we have called the Dawson range, lay a fine
glacier, sweeping round a curve of the valley. Its
surface showed a perfect labyrinth of crevasses, and its
sides were piled with avalanche *débris*. One avalanche
fell from a hanging glacier on Mount Fox while we
were looking on, and the fragments rolled far out on
the glacier below. By this time we were well into
the ice-fall, and as its worst portion still lay below us,
we paused. It was no road for the sledge, that was
certain. As a glacier pass it was quite practicable,

---

[1] Mounts Fox and Donkin we named in memory of the two members
of the Alpine Club who, with their Swiss guides, perished in the Cau-
casus while we were in the Selkirks.

but in the present state of its snow-covering, very dangerous for a party of less than four men well used to the rope.   Our two companions had never been on a glacier before, so taking this into consideration, and also the fact that, as would not be the case in similar undertakings in Switzerland, we should all be loaded with heavy packs, we reluctantly came to the determination that the descent by this route must be given up. We did not at all relish the idea of retreat, for this grand forest-clad valley, with its streams and glaciers, was quite unknown. Where did its river flow to? that and many other questions had for the present to rest unsolved. The glacier we had seen now took the name, on my plane table sheets, of the Geikie glacier, and we hoped some day to explore it more thoroughly.

It was near midday before we had rejoined the sledge and our companions. The sun had softened the snow, and the trudge up hill was fatiguing, for we sank knee-deep at every step. The question now was, should we explore a route down the glacier which poured from where we stood into Beaver Creek ? But as this would not give us much topographic information, we decided that observations from a knob of rock on the margin of the great snow-field overhanging Beaver Creek would be more useful, so taking up the ropes we started with our sledge in that direction. The sledge sank deep and pulled very heavily, we there-

fore took from it provisions for a meal, and started at
our best pace, leaving the sledge behind us.

A very gentle ascent led upwards to the bare patches
of rock, which here and there cropped up through the
margin of the snow-field, and at 4 P.M. we reached the
most conspicuous of these. The view was supremely
grand. Down below great precipices, capped for several
miles by a vertical wall of blue and white ice, lay the
verdant valley of the Beaver with its silvery streams.
The curved plateaux of the hills beyond looked even
more remarkable than when we saw them from the
shoulder of Sir Donald, and their origin now suggested
itself. The plateaux were most noticeable on the top
of the nearest range, that forming the opposite side of
the valley. The cliffs rose vertically from the Beaver
valley, above heaps of *debris* covered with forest. The
plateaux were separated from one another by low
nearly equidistant ridges at right angles to these
cliffs, and the whole surface seemed covered with
grass. On the north side of the low separating ridges
were thin lines of snow, not as yet melted by the
summer sun. Standing as we were on the great
glacier field, and noticing the knobs and ridges which
bounded it, we could not avoid the conclusion that
before the scooping out of the Beaver valley by its
streams, glaciers had moved from where we stood with
even flow over the hill tops opposite, and left these
ridges as lateral moraine accumulations. The valley of

the Beaver, at right angles to the ancient drainage
lines, had no doubt been scooped out by its stream
since the passing away of the great Ice age.

This work is still manifestly going on, for when we
peered over the brink of the precipices below us, what
devastation was manifest! Piles of freshly fallen rocks,
bare cracked surfaces from which they had fallen, loose
tottering blocks, and pinnacles of harder material
standing up from the rock buttresses; owing their
present stability to their hardness, but even many of
these looked just ready to fall. The remains of ice
avalanches lay in heaps, thousands of feet below us, the
white ice-blocks with lovely blue shadows being most conspicuous far down, almost to where the dark forest was
struggling upwards over the chaos of denudation. To
sit there and read the story of the hills and the valleys,
was as fascinating an occupation as I suppose could exercise the human mind, but we were thirsty and hungry,
and the slow process of melting snow on flat stones,
and catching the drops in our drinking cups, claimed
for some time our undivided attention. Having satisfied the cravings of nature we felt more inclined to
tackle to our work again. The panorama of the
Rockies was particularly interesting, and we took
bearings of the principal groups of high peaks. In
the direction where Mounts Hooker and Brown ought
to lie, there were no specially high mountains, the
whole range as far as eye could see, probably 200

miles from north to south, was snow-clad. The loftiest
groups did not rise much above the mean height.
Those we specially noted bore by compass 5° and 17°
west of north, and 55° east.

The afternoon was now advancing, and as we had
sketched and photographed the snow-field from every
possible point of view, we had to decide what direction
we should face for a camping ground. Up here was
no resting-place, no fuel, and it was doubtful if we
could follow the glacier down to the side of the mountain we had called after Professor Macoun. At first
we thought we might try the descent in this direction;
but calculating on the gain, which was uncertain, and
a very certain expenditure of energy in hauling the
sledge through soft snow, perhaps only to have to haul
it back again, we finally settled to fall back on
our old camping place on Perley rock, where at all
events we were sure of the fire-wood we had left
there. On regaining the sledge, we roped up, and for
some distance it came along well, for though the snow
was soft, the slope was in our favour. Then it
ploughed deeper and the hauling was most laborious.
It was only in spurts that we could all pull together.
Then a pack fell off, and we did not miss it till we
had travelled 500 yards and had to tramp back for it.
Clouds began to drift up in threatening masses, and from
distant valleys we heard the low growling of thunder.
This warned us to work hard if we wished to have dry

clothes for the night; and at last the little snow *arête*
was reached, and crossing it to the rocks, we lost no
time in setting up the tent over our bed of the night
before.

Speedily we lit the fire, as we hoped to cook the
marmot for supper. The setting sun threw a lurid
glare on the torn clouds, which were swirling round the
peak of Sir Donald. A few big drops fell; in an instant a fierce gust of wind swept down off the snow-fields,
making the tent flutter and flap to such a degree, that
we had to huddle into it, to prevent it from being blown
over the precipice. After a lull came another stinging
squall with a deluge of rain, and quite close to us, a
silvery white flash of lightning, darting athwart the
dark cliffs of Sir Donald, fairly dazzled us through the
tent walls. Crash came the thunder almost at the same
moment, blinding flashes followed in rapid succession,
but soon we were thankful to hear the thunder rumbling
away on the far side of the peak. The rain ceased,
we crept out of the tent, and relit the fire with wood
which we had put under the tent before the rain came
on; we had just time to fry some bacon and get a cup
of tea when the storm began again. Then we crept into
our sleeping bags and lay on the tent floor, which being
of one piece with the sides, made it impossible for the
tent to blow away with our weight inside. Covering up
our heads, we endeavoured to be as philosophical as
possible. When the elements near us were quiet

we tried to get to sleep, but none of us succeeded, till a steady deluge of rain and hail set in, and its monotonous sound drowning every other hushed us to repose.

When morning dawned the weather had cleared, and a keen breeze blew from the ice. We made a hasty breakfast, and packing up our things commenced the descent. We began by making a short glissade, those first at the foot of the slope catching the packs as they shot down from above. Then we climbed down some crags and reached the margin of the snow-covered glacier. The surface was frozen perfectly hard, and as I was first down, I commenced cutting steps across it. H. had delayed to rearrange his pack and was down last. Not noticing what I was at, and being a little hurried, he concluded that this snow was, as we had found it each time before, soft enough to kick our feet into. He accordingly leaped on to it without first testing it. Instantly his feet went from under him, and he fell heavily on his side and shot downwards. Instinctively he drove in the pick side of his axe, but he was going so fast that it ripped through the surface without stopping him. It was the work of a second, a glance told me it was all up if he could not turn his axe. Quick as thought he whipped it round, and the adze side held fast. Fortunately he was able to hold on to it, though the chuck, he said, nearly brought his arms from their sockets. We ought of course to have been

roped, but the little bit of glacier we had to cross was
so short we had not thought it worth while.

When I had reached safely the middle of the ice-
slope I attached the end of a long line, composed of
all our ropes tied together, to my axe, which I drove
down firmly into the snow. My companions then
fastened all the packs in one huge bundle to the
other end, and launching them off they first shot
down the slope, and then swung, like a pendulum, to
a place of security. Lower down, the snow was soft
enough to kick steps in, and in a few minutes more
we were on the boulders.

Drops of rain began to fall, and by 9 A.M. there was
a steady down-pour, accompanied by a strong north-
east wind. This continued during the whole time we
were descending. Our two companions caught a
glimpse of a bear at the foot of the glacier, but we
were separated from them at the time, and after a
short halt for dinner we passed our lower camp, and
reached the inn at four o'clock. It was Sunday, and
we had done rather more than a Sabbath-day's journey,
but were glad to conduct divine service for ourselves
and the few inmates of Glacier House in the evening.

CHAPTER VIII.

" But descending
From these imaginative heights, that yield
Far-stretching views into eternity
. . . . to Nature's humbler power.

.   .   .   .   .

Where on the labours of the happy throng
She smiles, including in her wide embrace
City, and town, and tower—and sea with ships."

WORDSWORTH.

We start for Vancouver.—The gorge of the Frazer.—The salmon
canneries.—Back to the mountains.

OUR faces not being in a comfortable state of repair,
the skin having peeled off from long exposure on
the snow, and the weather being still very threatening,
with thunderstorms wandering about the valleys; we
determined to make a three days' excursion to the
western end of the line—to Vancouver and back—and
so give our countenances a chance of healing, and the
weather of mending, before making any other expeditions
on the glaciers.

On July 31st we accordingly took our seats in the

west-bound train, and started down the valley of the
Illecellewaet.   As we swept round the curves of the
Loop we stood on the platform of the car, and got
new views of the mountains.   That up the valley
towards the glaciers of Mount Bonney was particularly
fine, and now for the first time we recognized that
the route to this massive mountain must be sought
for by ascending the Loop valley.   Passing Ross peak
station, we entered the forest-clad defile, crossing and
recrossing the torrent as the valley offered better
conditions for the railway.   At about fifteen miles
from Glacier, we stopped at Illecellewaet, a typical
frontier village, the inhabitants being all prospectors, miners engaged in the silver mines high up the
mountains to the northward, lumber men, and those
associated with the Canadian Pacific Railway.   Burnt
black trunks alternated with wooden houses, some
of which stood on legs in swampy pools only half
reclaimed from the overflow of the river by piles of
empty meat tins, broken packing cases, &c., which
were littered about everywhere.

After a short halt the breaks released their grasp,
and we sped onwards down the valley.   At Albert
cañon we halted to see from a stage the surging
waters of the Illecellewaet (*rushing water*), a very
appropriate name just at this place.   The main stream
was here joined by its north fork, and the combined
torrent foamed and surged through a deep cleft in

the mountain. Many of the passengers tried the experiment of throwing down stones, to test the common
belief that the blast of air near the water, is so great
that the stones never reach the torrent.   Then we sped
onward once more; at last the gradient decreased,
and we ran out into the flat wide valley of the
majestic Columbia.   On its further shore rose the
Gold range, crowned, above the forest-line, by snowy
peaks; and as we crossed the bridge and commenced
the ascent to the Eagle pass, we were able to look
back and enjoy lovely views of the Selkirks.   H. well
remembered his former experiences here; how, after
stumbling over fallen trees in the gorge of the Illecellewaet, and losing one horse, they here emerged
in sight of the haunts of men, and swam their horses
across the river.   The whole of the region is one of
stupendous forests, the cedars, Douglas firs, and other
coniferæ becoming larger and larger as we approach
the flat land near the coast.   In many places however these forests, far as eye could see, were represented by nothing but charred and blackened trunks,
relieved here and there by the verdure of the young
forest just starting up around the old stems.   Again
the young forest itself had been burned, and was now
no more than an impenetrable thicket of dead poles.
In the evening light we skirted for many miles the
placid waters of the Shushwap lake.   Flocks of ducks
rose from the edge of the track, and fluttering out

I

a little distance, broke for a moment the calm reflections of the mountains.

Leaving the lake by the valley of the Thompson, we
passed about midnight through the town of Kamloops,
an old-established settlement of the Hudson's Bay Company. Most of us were asleep by this time. When
I awoke it was broad daylight, and the scene presented
a great contrast to the views on Shushwap lake, with
its calm mirror-like surface and sombre woods. Now
we were descending a wildly savage defile. Bare
reddish rocks and slopes of loose *débris* flanked its
sides. Over these slopes the train had to go at
a cautiously slow pace, for we could see the stones
rattling down as we passed. Deep down,
hundreds of feet below the track, the Thompson
roared in wild rapids over its rocky bed. Descending rapidly till the line got nearer to the river, we
reached the junction of the Thompson and the Frazer,
and then, for the rest of our journey, we followed
the latter stream. The scenery in the cañon of the
Frazer is almost the finest on the whole journey, and
we congratulated ourselves on not having returned
from the Far West without seeing it. For miles the
track is laid on ledges or on trestle supports against the
face of the rock, or in cuttings and through tunnels at a
height of about 150 feet above the foaming river. The
jutting rocks were in many places crowned by stages,
where the Indians dry their salmon, and from some

of these numbers of salmon with bright red flesh were
hanging to dry. Over the swirling eddies of the
river, Indians might every now and then be seen
crouched on a scaffold of long poles, suspended by
ropes from the overhanging rocks, and patiently watching their chance of a fish entering the huge landing
net, which they plunged apparently at random into
the eddy.

This splendid river, 1000 miles long, was discovered
in 1793 by Alexander Mackenzie, after whom the
Mackenzie river is named. He was a partner in the
North-West Company. He struck the head waters of
the Frazer by crossing the divide from the Peace river,
but believed it to be the Columbia. Thirteen years
later Simon Frazer followed the river from its source
to the sea, and gave to it his own name. The
story of his canoe voyage is one of thrilling adventure,
and now as we looked down upon the foaming rapids,
we could realise to some small extent the awful risk
of shooting downwards, in a heavily laden canoe of
birch bark, when all ahead was quite unknown. Frazer
and his Canadian voyageurs, in accomplishing this
task, well earned their place in the history of exploration.

High up on the opposite side of the cañon the
Cariboo trail was visible, but is now almost deserted in
favour of the railway. Along this narrow road the
teams used to bear the rich produce of the best mining

district in British Columbia, and along it many a
poor fellow has trudged with his pick on his shoulder
and bright hopes in his heart, fulfilled to the few,
but resulting in disappointment to the many, in the
great hunt for gold.

Of what wonderful use the search for gold has been
in the development and civilization of the world!
Any one thinking over the history of the new world
of Australasia or America cannot fail to be struck
by the fact, that in the economy of Nature gold
has proved, before all things else, the great incentive
to exploration. The anxiety to obtain it has set on
foot those projects of discovery, which might never
have started had it not been for this stimulus.

The Spanish conquest of America, the seeking for
El Dorado, the adventures of Raleigh, the advance
of Australia, the opening up of California, explorations of British Columbia, and the advance from the
known regions into the unexplored, all turn on the
one thought—gold.

And then when gold has done its work, when the
new country is known to the world, when the waste
lands are reclaimed, when prosperous communities are
established, then it sinks into insignificance compared
with the other industries, the prosecution of which
lead to civilization and a better ordered society. A
country so difficult to travel in as British Columbia,
might long have been, for all practical purposes, a

terra incognita were it not for those pioneers of civilization, the prospector and the miner.

It was on the Frazer river, in the year 1857, that gold was first discovered in British Columbia, and many places we passed were of historic interest. The great rush to the Frazer took place the following year. In 1861 extensive discoveries of gold were made by prospectors in the Cariboo district to the northward. The lower Frazer soon became deserted for the richer placers of William's Creek and other sites in Cariboo. The difficulty of getting there *viâ* the Harrison river was so great—the distance from New Westminster, whence all supplies must be drawn being 520 miles, that prices went up to a marvellous extent. Men's wages rose to fifteen dollars per day, and needles were sold for one dollar each. One man, named Ned Connel, did a "good thing" in beef. Buying an ox at Lillooet, where the pack trains for Cariboo used to load up, he trained it to carry a pack. His was the first pack animal to reach William's Creek. Ned made well on the freight of the goods he carried; then he killed the ox, sold the beef, 900 lbs. in all, at sixty cents. per pound, thus realising over 100*l.* net profit on the one beast.

A great number of Companies work in Cariboo. The names of some are suggestive, and possibly investors may like to buy shares in the "Never Sweat Company," or, for contrast, the "Wake-up-Jake Company." The probability however in the case of such investments is that they are allotted claims in "Humbug Creek." At the present moment Cariboo is in "full swing," and certain gentlemen, who ought ere now to have experienced similar conditions in a literal sense, infest the road, and "hold up" the coach every now and then. By terrorizing the driver and passengers with their Winchesters, they take what they want in the way of gold, and then let the passengers, as a rule, depart in peace.

Though gold has had such a place in the development of the country, the first explorations of British Columbia were undertaken for the sake of furs. The North-West Company, the great rival to the Hudson's Bay Company, established themselves in British Columbia early in the present century.

Getting distant views of the fine peaks of the Cascade range as we leave the gorge of the Frazer, the line traverses some flat, heavily forested country, where the pines grow larger than any we had yet seen, and the half deserted Port Moody is reached. The waters of the Pacific ocean lap the railway, and at 1.30, or 13.30 as that hour is called here, we run into the station at Vancouver, the terminus of the line. The steamer from Yokohama was lying at the wharf discharging tea, and we learned that fifty per cent. of the whole tea trade of the continent comes by this route and finds its way to the more populous

side of America, and even to Europe, by the Canadian
Pacific Railroad. Vancouver, only two years old, has
a population of 20,000; this was one of the chief
facts of interest about the place. The heart of the
town was devoid of houses, owing to the lands being
held for higher prices. In the streets already built,
every second house seemed an office for transfer of
"real estate," speculation in building sites being the
one thought of paramount importance in Vancouver.

Beyond these facts few places in the world could
surpass Vancouver in want of interest, and as I was
particularly anxious to see something of the great
salmon business at New Westminster, we left Vancouver about 5 P.M. for that place. New Westminster
has the air about it of an old settled town, and
the people seem to belong to the country, and are not
like the unrooted exotic slips stuck down in Vancouver.
The inn was most comfortable, thoroughly "Colonial,"
as its name indicated; meals well served, and for
dinner, bed, and breakfast, we only paid one dollar
each.

Early next morning we set off, under the
friendly guidance of one of the leading citizens, to
salmon cannery, where we saw the Indians bringing
in salmon in their canoes; Chinamen splitting, clean-
ing, and packing, and Europeans overseeing the work.
Most of these salmon are taken in drift-nets im-
ported from Glasgow and the north of Ireland, the

cedar-wood floats, instead of corks, alone giving the
nets a local character. The soldering of the tins, in
which the fish was packed, was one of the most
interesting points in the whole process. The cans,
with the lids just squeezed on, were caused to roll
along an incline on which a trough of solder was
kept liquid by a furnace beneath. Through the
melting solder they rolled, tilted on one edge, and
then went on their way down the incline, which was
just long enough to give them time to cool.

The salmon taken in greatest quantities in the Frazer,
is the blue-back, *Oncorhyncus nerka*. The flesh, like
that of the *quinat* of the Columbia and other Western
rivers, is of a deep red colour, and not so rich as that
of the *Salmo salar* of the Atlantic basin. Unlike the
latter fish, these salmon readily take a spoon-bait in
the salt water, but once they are in the rivers they
will look at neither bait nor fly. On the Columbia
there are great engines driven by water-wheels for
ladling the fish out of the river wholesale, and the
amount consumed each year, and sent across the
continent from the Pacific coast rivers amounts to
thousands of tons. About 2,500 tons of canned salmon
is a fair estimate of the annual out-put of the canneries
of the Frazer river alone. Whole trains freighted with
canned salmon cross the Rocky Mountains, and the
bulk of the production of the British Columbian rivers
seems to find its way to England.

All the houses in British Columbia are roofed with wooden shingles, and as we left the salmon cannery we passed through a lumber yard where steam saws were cutting shingles and filling the air with the delicious perfume of the fresh cut cedar. Before our train left New Westminster to meet the east-bound Express from Vancouver, I had just time to call on an old acquaintance, the worthy Archdeacon Woods of New Westminster; then getting on board the train, we reached the junction, where while waiting for the Express we were nearly devoured by mosquitoes: at last our train arrived, we took our seats, and as the shades of evening closed in we found ourselves once more in the cañon of the Frazer, *en route* for the snows of the Selkirks.

Early next morning we were following the course of the swift Thompson upwards, and were much interested in the wonderful development of the bench and terrace formation in the bluffs of silt, but sparsely covered with vegetation, which bounded its course. All seemed going on well till we reached Kamloops, about breakfast-time. Here the guard informed us we must expect some delay, as the west-bound train was off the track about a mile from the station. We walked along the line to the scene of the accident, meeting as we went many of the passengers on their way to Kamloops to get breakfast. A heavy thunder-shower had washed down a quantity

of gravel over the rails, and in the darkness of night
the west-bound engine had mounted the heap of *débris*,
and then missing the rails pulled half the train off
the track, which was now buried to the axles in the
loose soil. Slowly the section men raised the locomotive inch by inch with screw-jacks. The tender
gave much more trouble, then the cars were hauled
on by a spare locomotive, and after nine hours' delay
the track was cleared and we were able to pass, but
the rails were so twisted and loose that the chances
of the wheels of our train slipping off seemed about
ten to one. As we entered the Selkirks the weather
was dark and threatening. We could see nothing
of the mountains, and we reached Glacier House at
10 P.M. in teeming rain.

CHAPTER IX.

" To regions haste,
Whose shades have never felt the encroaching axe
Or soil endured a transfer in the mart
Of dire rapacity."

WORDSWORTH.

Pioneering the Asulkan pass.—Rocky Mountain goats.—The Dawson
range.—Reconnaissance of the Loop valley.

THE acquaintance we had made with the mountains round " Glacier " had enabled us to form a fair picture of the region we were going to explore. From among all kinds of possibilities which had suggested themselves on our first arrival we could now select those which would be most helpful to our main undertaking. Foremost amongst these were the exploration of the unknown valley (to the southward) which we had seen; and partly in connection with this, and partly for interest on its own account, would be the ascent if possible of Mount Bonney. Having learned by our late experiences that it was useless to start heavily laden with packs until we had pioneered the route before us, we determined to start as lightly equipped as possible on an excursion up the valley towards the glacier col which we have called the Asulkan pass, and see what it would lead to, and also to inspect the valley opening at the Loop with a view to getting at Mount Bonney.

After breakfast, at 7 A.M., we started in company with the good dog "Jeff" for the first of these excursions, taking nothing with us but the two cameras, small plane table, rope, axes, and sandwiches. Clouds were still hanging about the mountains and the vegetation was dripping after the recent rain. Several times during the day we were as wet as though we had fallen into the river, but dried quickly again under the breeze and sunshine.

On leaving Glacier we bore away to the right, keeping to the path which had been made to give access to the valley. For about a mile, it was in good order; then we found it obstructed by fallen trees, and soon after it vanished. Following the eastern bank of the torrent for some distance we crossed to the other side on a fallen tree. Farther on we recrossed on a bed of consolidated snow which completely bridged the torrent. Then we traversed a fine tract of coarse meadow land, which ought to have had about fifty head of cattle on it. Steep cliffs, down which numerous waterfalls splashed from the snowfields above, re-echoed the music of streams. Dark

pine forest contrasted with the bright green grass
levels below and the sunlit glaciers above. It seemed
just like some scene in the Engadine, and was in itself
a perfect gem of an Alpine valley.

Passing these meadows we returned to the gravel flats
near the river, and sought for means to cross to the
eastern side once more. A fallen tree lay in the
surging waters, part of its trunk under water. With
difficulty we made the first step over a deep swirl
to this trembling, quivering foot-hold. It was a
bridge more fitted for a rope-dancer than for us.
Jeff would have none of it. He preferred swimming
the torrent, and was consequently rolled over and
over, and carried two hundred yards down the stream
before he reached the farther shore. Keeping to the
level swampy land as far as possible, we were soon
stopped by alder scrub which forced us up on to the
forest-clad mountain side. We halted for a moment,
to admire a lovely waterfall coming down several
hundred feet in one leap from the opposite cliffs.
Then for an hour we had to struggle through a maze of
dense forest and fallen logs, followed by a scramble
through the alder scrub lining the cañon,
which the glacier torrent had excavated. At
last we were clear of vegetation, and crossing the
stream for the fourth time, on a snow bridge, we
commenced to ascend the high moraine to the glacier.
This moraine, like many a one in Switzerland, was

composed of semi-rounded boulders compacted together
by pulverized rock, channeled by watercourses, the
lower part a perfect garden of flowers, while the
upper edge, recently formed, held stones barely poised,
and ready to roll down at the first touch.   The
flowers, among which our course now lay seemed
wonderfully brilliant in colour.   The scarlet castalleia,
the purple-flowered epilobium, yellow asters, and
white saxifrages bloomed in masses or in long lines
following the ridges not recently disturbed by the
last snow-melting. When we came to the steepest
slopes, all vegetation ceased, a continued crumbling
down was going on, and we had to beware of following in each other's tracks too closely, owing to
the showers of boulders which became detached
from the mass as we climbed.   As for Jeff he
would now and then make a most reckless charge
ahead, sending down a perfect cannonade of small
stones.   The moraine, where we climbed it, was about
300 feet high, and on gaining the top we saw that
it would not do for us to follow it, but that we
must cross the glacier. This we did without much
difficulty below the ice-fall, and then ascended
the eastern moraine, at the side of the broken
seracs.

Above the ice-fall the moraine soon came to a
termination, and putting on the rope we took to
the snow-covered glacier.   Here and there crevasses

made themselves apparent as mere dimples in the
spotless surface; others were wide open, and we
were much amused at the sagacity of Jeff. He
sprang across the open crevasses in the most reckless
manner imaginable; but when he came to one only
six inches wide, he whined most timidly, and peered
down anxiously into the darkness. I don't know
whether he had some intuitive feeling of danger, or
whether in our greater caution he recognized its
presence, any way he showed most distinct evidence
of alarm, and relief when the danger was passed.

We soon got clear of crevasses, and after a trudge
up a gentle incline, reached the col at 2 P.M.
The scene was of course most interesting—deep
down before us lay the valley we had looked into
from the great snowfield. Our view was now across
it at right angles. Some snow slopes flanked our
col to the southward. Down these we quickly
ran, and gaining a projecting knob of rock improved
our view considerably. The Geikie glacier, with a
most wonderfully fissured surface, lay far below in
the valley's gloom. From the Dawson range right
opposite to us a most typical glacier, with lateral
and medial moraines descended and just stopped short
of being a tributary. Having taken several photographs
and the bearings of the principal peaks, we sat down
for lunch.

It was a lovely day, and we planned that this

certainly was a route to be adopted for future exploration.

While eating and making these plans, we saw two white specks moving on the grass slopes below, which as they approached, we soon made out to be wild goats. Six others joined the first two, and came up to have a look at us, and then grazed without showing the slightest alarm. The long white wool hanging thick above the knee gave them the appearance of wearing knickerbockers. As we had no rifle they were perfectly safe, and "Jeff," who was death on all kinds of small game, sat on his tail and looked at them with much complacency. They had never seen man or dog before, and "Jeff" had never seen that kind of beast except in the midst of civilization, and being a civilized dog he felt that barking at, or hunting goats would be the lowest depth of depravity.

Other thoughts passed through our minds, for we saw in them a future store of good food when we should come camping into this valley.

There was now no time to make further explorations, so taking a last look at the goats we called the pass by their Shushwap Indian name, *Asulkan*, and picking up our light swags ascended to the col, and then trudged down the gently sloping glacier for a couple of miles at a swinging pace. When the few crevasses were past and the moraine reached, we threw off the rope which, except as a matter of principle, we might have done very well without. Following our former track in every detail, we were soon in the forest—then came the swollen river which had to be crossed on the fallen tree. We put up a brace of snipe in the swamp near it, and observed some little sandpipers on the gravel spits; then once more we were in the forest, and following the course of the river reached Glacier House at 8 P.M. after a delightful and most profitable day of twelve hours.

We spoke to Mr. Bell Smith of the beauty of the valley we had been in. He too had done a good day's work, and told us that while sitting at his easel in a secluded part of the forest, a bear came and looked at him, and grunting approval of his occupation, went on his way in peace.

One future route had now been explored, so our next move was a reconnaissance of the Loop valley and a route to the foot of Mount Bonney.

After an early breakfast, taking our axes, a prismatic compass, and small detective camera, we started down the railway track to the westward to where, on immense trestle-bridges, the line forms a double loop like the letter S.

On reaching the first high trestle-bridge, beneath which the turbid glacier torrent from the snows of Mount Bonney, finds its way to the Illecellewaet, we

K

paused to consider which side of the stream we had
better try. The whole valley was clad in trackless
forest, and so far as we could learn, no human being
had as yet penetrated to its head. As the forest
looked much the same on either hand, but as the
mountain side beneath Ross peak looked the most
precipitous, we decided on the more gently sloping
right bank.

Divesting ourselves of our coats, which we placed
on a conspicuous fallen tree until our return, we
entered the forest, fully prepared for a hot and hard
struggle. It is difficult to give anything like an
adequate idea of what such forests as these are like.
Besides the noble pines in the prime of life, dressed
with lichens, the young trees growing up, the
thickets of blueberry bushes, rhododendrons and the
devil's club with its long broad leaves and coral red
fruit, but most terrible thorns; there is the network of
fallen trees, some rotting on the ground, others piled
on top of these at every possible angle, with stumps
of broken branches sticking out like spikes. Again,
overhead are trees recently fallen, jammed against
others, some only needing a push to bring them
down. Getting through such a tangle is all hand
and knee work. Sometimes a fallen log leads in the
right direction, and you can walk along it, if the
rotten bark does not give way and deposit you in a
bed of devil's club. A few hours of this kind of

work is a desperate trial to one's temper, you make
so little progress for all the labour expended.

On this day, three hours of it brought us just
to the bend of the valley, and we saw plainly enough
that this was the route that must be followed if we
would reach Mount Bonney.   Now we had no packs
—what would it be when we had!   But there, in
full view, was the grand glacier heading the end of the
valley : it could be reached with fine weather, patience,
and perseverance.   We hoped we might be blessed
with all three.

CHAPTER X.

" In the calm darkness of the moonless nights,
  In the lone glare of day, the snows descend
  Upon that mountain ; none beholds them there,
  Nor when the flakes burn in the sinking sun,
  Or the star-beams dart through them. Winds contend
  Silently there, and heap the snow, ...
  And what were thou and earth and stars and sea
  If to the human mind's imaginings
  Silence and solitude were vacancy ? "

                        SHELLEY.

Start for camp in Loop valley.—The Glaciers of Mount Bonney.—
        Ascend the Lily glacier.

ON August 6th we started with our small tent, blankets, rifle, instruments, &c., for the Loop valley, determined if necessary to spend the week in Mount Bonney's conquest. We had an early breakfast, and Mr. Perley kindly offered to help us so far as he could. A small truck fitted to run on the railway, provided with a lever break, was ready to hand. We lifted it on to the rails, packed our goods on it, and sitting on the packs with Mr. Perley and two men, who volunteered to shove the truck up the hill again after we had started, sped away down the gradient, round curves and through snow-sheds in quick succession, and crossing the first high trestle-bridge of the Loop, brought our truck to a stand at its farther end. From our experiences in preliminary exploration, we judged that nothing could be worse than the forest on the right bank of the creek, so we decided to make our way up the left bank and take chance for its being better. A patch of flat shingle in the river-bed also gave us a clear start of about 300 yards, so lifting our packs off the truck, we pitched them over the embankment, and saw them bounding and hopping down to the bed of the creek a hundred feet below. Bidding adieu to our friends, we left them to shove the empty truck up the gradients towards the inn, and following our packs down the embankment, commenced work in earnest. There were four packs to carry, but as they seemed too heavy, we took some tins and cartridges out of them and made a *cache* near the river bank. Then taking a load each we soon reached the end of the shingle flat, and were forced by the stream, swollen with melting snow, into the forest. Setting down the first packs, we returned for the others and then moved on again. To carry them continuously on our backs was impossible; for the obstacles—composed of fallen trees, great boulder heaps shot down from the mountain precipices above, or alder scrub which generally grew densely where

the forest had been smashed down by snow-slides—
were so difficult to overcome, that tumbling the pack
down a slope, or pulling it up after us when we had
gained firm foot-hold on a log, was the easiest way
of getting along.   Having two packs each neces-
sitated our going over all the ground twice.   When
four hours had been thus spent, and the shingle
of the river bed was once more reached, we felt
pretty tired, and were glad to halt for dinner.   The
way ahead was now more open, and after a good
rest we set to work again.   The loose, angular
boulders over which our path lay were a delightful
change after the forest.   Too soon, unfortunately, this
charming experience came to an end, and once more
alder scrub and forest blocked the way.   Below on
our left the creek was arched over with snow thickly
strewn with trunks of cedars and hemlocks brought
down by spring avalanches, and for more than a mile
the river was completely invisible.   The surface of
this snow seemed the most promising route for us
now, and descending the precipitous bank we gained
its surface.   The tree-trunks even here were a serious
obstacle; but we followed up the river bed for half a
mile.   It was now high time to think of camping for
the night, but not a single level spot could be found.
We set down the packs and climbed the hill-side
to explore : nothing like a camping-ground was
visible.   Our little tent needed very small accommo-

dation, but not even that could be found. It was wonderful to see the way that great cedars, four feet and more in diameter, had recently been snapped across by the avalanche, the unmelted snow of which formed the covering over the stream. Huge trees were split from end to end, and the *débris* formed a savage foreground to the swelling slopes of forest, the blue ice of the glaciers filling the head of the valley, and the dark purple precipices of Mount Bonney, crowned with a cornice of snow sharply defined against a cloudless sky. The banks of the torrent, above its snow covering, were here and there gay with bright flowers, scarlet and purple being the predominating colours. The air was loaded with the fragrance of the pine forest, and above the hoarse roar of the torrent and the splash of a cascade, which cut its way through a rocky cañon on our right, the shrill cry of the hoary marmot sounded weird and startling. Having given ourselves a few minutes to drink in the magnificent beauty of our surroundings, and to discuss the two possible lines of attack on Mount Bonney, evening tints warned us that there was but little time to lose, and as it was too late to experiment on the unknown difficulties ahead, there was nothing for it but to trudge back with our packs to where we had taken to the snow-choked river bed, re-ascend the steep bank, and camp on the boulders close to the shelter of the forest. We noted the place when passing it on the way up, but on regaining it things did not look so promising; nevertheless, there was no choice—that much was certain; so by building up and levelling down, turning up the smooth side of the stones, and then making a bed of pine tips, we pitched our tent on the top, and arranging a few stones for a fireplace at the tent door, soon felt quite at home, and ready to invite a friend to dinner should one turn up, which, however, was not likely.

The sputtering of the bacon in the frying-pan, the jet of steam from the kettle, the glare of the cedar-wood fire, soon made us feel at peace with all the world; and after sitting for a while to enjoy the music of the torrent as it roared over its rocky bed, carrying boulders along, the clinking of which one against the other was distinctly audible, we turned into our sleeping-bags; for though these valleys were intensely hot in the daytime, the downward draught from the glaciers which sets in immediately after sunset, makes the air icy cold.

Our intention for the morrow was to explore the pass to the east of Mount Bonney by ascending the glacier which I have named on my map the Lily glacier. We examined the route to it from a little above our camp before night closed in, and hoped that it would give us a further view into that valley which we had already reached over the Asulkan pass. Perhaps, too, we might find the far side of

Mount Bonney more accessible than the side facing us, if so, we might try it. When a place is quite unknown, plans cannot of course be made with any greater certainty than this, so we went to sleep with that delightful uncertainty as to the future, which lends such a charm to travel; except when it is an uncertainty involving a prospect of short commons.

At 4 A.M. we were awake, and after a hasty breakfast made ready to start on our exploration of the Lily glacier, and the pass at its head. The rifle, the camera, plane table, &c., and provisions for the day were divided as fairly as possible between us; and taking our axes we descended to the snow-covered river bed, and followed it upwards over the fallen tree trunks which strewed its surface, till we reached the upper termination of the snow bridge, and then we had to take to the forest on our left. Now our difficulties commenced in earnest, and for the next two hours I thought several times of giving up in despair.

The river bed contracted to a narrow cañon, one side being composed of rotten slopes of *débris* produced by the disintegration of layers of mica schist which cropped out nearly horizontal. The opposite side of the chasm was strangely enough composed of great vertical slabs of quartzite, against the base of which the torrent swirled along.

This rotten slope on which we were, had here

and there crumbled away into perfectly inaccessible declivities; now and then we cut steps with our ice-axes, but the greater part of the time was spent in forcing our way upwards through alder scrub, trusting almost entirely to our arms and having no foot-hold often for many yards. Not knowing what difficulties might be in the forest above us, we tried to keep as near as possible to the river. The distance to the foot of the glacier with a fine ice cave from which the torrent issued, seemed absurdly short; but there we were—scrambling up a slope 100 feet in order to get past some difficulty not ten feet wide. Then down again to the boulders of the river, then up a couple of hundred feet, till at last we determined to give up the route by the river and strike straight up for the pine forest above. Ascending by a slope of ochrey yellow detritus we got up under the over-hanging roots of the hemlocks and balsams, and crept along beneath them for some distance. Then being forced to descend about fifty feet we at last found a place where we could clamber up by roots and so gain the firm slope above. We felt relieved at being done with the crumbling schist, but a chaos of fallen trunks made lateral advance still impossible —there was nothing for it but to go higher. And when about 500 feet above the stream, what was our surprise to meet with a fair lead along what was most distinctly the top of an old glacier moraine.

The moss-covered boulders were arranged in the
most symmetrical manner possible, and the depression
in the upper side afforded pleasant walking, whenever it was not encumbered by fallen logs. If proof
of the shrinking of the glaciers in these regions was
wanting here it was. Nothing could be more patent
than that at some period not very remote the glaciers
of Mount Bonney filled this valley to a depth of
500 feet greater than at present. We followed the
moraine through the forest for about half a mile,
but then it terminated, with the forest, at the margin
of an avalanche track overgrown by alder scrub.
Half an hour's scrambling through this brought us at
last clear of vegetation into the hollow between the
glacier and the mountain side. The recent moraine stood
up on our right capped by the clear ice of the glacier,
from which stones occasionally rattled down, but the
snow-filled *langthal*, or depression between the mountain
and the moraine, was filled with compact snow which
promised us easy going for a mile or so. We
found that a large bear had just taken to the same
route in advance of us so we looked to our rifle and
followed his tracks towards the pass.

The moraine on our right was that of the main
glacier coming from Mount Bonney ; it now diverged
more from the mountain, and the snow we had
found so pleasant to walk on terminated in a waste
of boulders through which a muddy torrent flowed

from the lower end of the Lily glacier, which now
blocked the way ahead. The mountain side to our
left was still clothed in rank sub-alpine vegetation,
the large succulent-leaved *vcratrum viride* affording
dense cover for the marmots and other creatures
uttering shrill whistling cries of alarm. The
struggle through the forest had taken more out
of us than twice the same amount of work
on the open mountain side, so before commencing
further difficulties, which the steep termination of
the glacier now presented to us, we halted at the
clearest of the streams to rest and have something
to eat. The glacier before us was quite inaccessible
for some distance, stones falling from the lower face
warned us to give it a wide berth and, as far as we
could see, it was broken up into spires and pinnacles
of ice. Leaving it for the present on our left we ascended the high moraine separating it from the Mount
Bonney glaciers, and the view over these from its summit
was very fine indeed. This great field of undulating
and in some cases much broken ice, divided itself
into no less than seven distinct glaciers, separated by
moraines ; beyond these, in a hollow of its own, was
the glacier descending from the col I have called
Ross pass, and this with the Lily glacier on our left
made nine glaciers all converging to the narrow cañon
through which we had attempted to make our way,
and through which the combined streams roared

and swirled. It was a grand amphitheatre paved with ice, walled in by the dark precipices of Mount Bonney, Ross Peak, and the high ridges facing them, and presenting a comparatively small opening towards the Illecellewaet valley. Before going further I took two photographs, and then we resumed our journey, the easiest route being the very narrow top of the moraine. Here we found bits of vein quartz with bright cubes of galena, showing that somewhere in the ridge bounding the pass to the south-westward a vein of this mineral may be discovered. Following the moraine upwards for about half an hour we reached the mountain spur. The Lily glacier on our left was yet too much broken for us to take to it, so we ascended the *arête* which got higher and higher above the glacier. The rocks were split up and piled in wild confusion, some great blocks ready to start downwards on the slightest push. Now the glacier below us seemed more level, so although a good many crevasses, partially covered with snow, were visible, we took the first available chance of descending to its surface. This was done by scrambling down a remarkable natural tunnel through the rock, and then adjusting the rope we stepped on to the ice.

For a few hundred yards we had to observe the greatest possible caution, as the giving way of a snow bridge would be most serious with only two on the

rope. It may be thought that under such circumstances it would be better not to be roped at all.
This however we proved on several occasions to be
quite incorrect. Short of the collapse of a snow bridge,
the possibility of which we were most careful to avoid,
there are numbers of small slips, sure to occur, when
one can readily help the other, but which without the
rope might easily become dangerous. The crevasses
were soon passed, and we faced up the gentle slopes
to the col. The weather was not improving; thin
films of mist began to drift round the higher ridges,
hanging in the gullies and bringing out the details of
every crag which are quite invisible in bright sunshine.
When we reached the summit of the pass, it was past mid-
day, so all thoughts of attempting Mount Bonney by
this route left our minds. So far as we could see
up the rocks, the route seemed possible but not easy;
we knew the distance from the pass over an outlying
peak, down into a hollow, and up to the final summit
was very long; it was therefore too serious a piece of
business to undertake so late in the day. The view
from the pass was somewhat confined by a great ridge
from Mount Bonney running down towards the south,
and on the other hand by the ridge separating us from
the Asulkan valley; but what we could see, in spite of
the clouds which hung low, was most interesting.
The unknown valley that we had already got glimpses
of from the great snow field, and from the Asulkan

pass was again before us, but from a new point of view; and now for the first time we could see up a large branch valley, which was headed by a fine glacier that I have named after Sir W. C. Van Horne. This glacier was most symmetrical in shape, its moraines well developed, and the extensive snow-field from which it descended culminated in a perfectly spotless snowy peak.

The stream from this glacier was soon joined by others from the smaller glaciers which filled the lateral valleys, and with many a curve, it flowed through green alder scrub, to join the stream from the Geikie glacier, a little to the right of the point to which our view extended. Until now we had thought that this valley, flanked with fine pine forests, was the outlet of the drainage of this region, now we saw that it brought a fine tributary to the main stream. The outlet was still undiscovered. Continuing our course beyond the pass we commenced to descend the glacier before us, and bearing off to the south-east traversed some domes of snow which overhung the valley and promised a further view. We could trace the course of the river a little further, and conjectured that it must give a sharp turn to the southward; but after a halt for dinner and photography we returned to the col more anxious than ever to reach the summit of Mount Bonney, from which alone all these problems could satisfactorily be solved.

On the pass we set up the plane table. The backward view enabled us to mark points already fixed, and the forward glance gave us bearings of points some of which we had observed from former stations, and of others completely new. We had to work quickly, as the clouds were closing up, and it seemed certain that ere we could reach camp bad weather of some kind would be upon us.

Fortunately for the plane table work it was still quite calm. As we descended the glacier however the downward blast from the pass became stronger, and when we had descended about 500 feet we stopped to take a reading of the thermometer, and found, as I expected, that the temperature was eight degrees lower than at the summit of the pass. Further down it felt still colder. Our tracks were quite visible till we came to a steep part of the glacier, where the snow was blown off the ice and numerous crevasses stood wide open. We here missed our former track, and therefore experienced some delay in finding a safe route, but finally reached the natural gateway in the cliffs, and ascending through it quitted the Lily glacier. Then came the descent along the top of the moraine, and just as we reached the compacted snow in the hollow between it and the mountain side it began to rain. The snow gave us good travelling, so we lost no time and soon reached its lower termination. The ice of the main glacier from Mount Bonney had here broken down the moraine, and some crevasses formed regular ice caves, easy of access. Not wishing to get our clothes wet, as we had no way of drying them and needed them to sleep in at night, we proposed to take shelter in one of these ice caves and give the weather a chance to clear. We were of course aware of the danger of stones falling from the ice above, so no doubt the idea had in it a total lack of the prudence which I have always claimed for myself as a special characteristic. This time however we got our lesson. We had just diverged from our track and were making our way over some *débris* to get to the cave, when crash! down came about ten tons of rocks and ice from the glacier above right across its mouth. If we had been ten yards further! This thought flashed through our minds simultaneously, but I am afraid that what was expressed on our countenances when our eyes met was, "What an awful fool you were to think of going near that cave." This is one of the great advantages of two travelling together, that each can always have the luxury of saying when any mistake is made that it was altogether the other's fault. All the same, it was necessary to keep my jacket dry, so instead of turning it inside out, as an Irishman is generally supposed to do under such circumstances, I took it off and packed it tightly into my knapsack.

We left the glaciers and struck into the forest. The alder bushes were so dripping that we were soon just as wet as if we had been chin-deep in the river.

L

Wishing to avoid the difficulties of our ascent we took
a high level through the tall timber, found the ancient
moraine in the forest, and followed it till we were
immediately over our camp, which we could recognize on
the opposite side of the valley.   The rain, for the most
part a drizzle, now and then fell in a perfect torrent,
compelling us to take shelter under a large pine tree
from its chilling influence.   Then came the descent—
down for hundreds of feet by half-hidden boulders
strewn with fallen logs; down through the rhodo-
dendron bushes; down through blueberry bushes laden
with rich fruit, which of course we could not pass by
without picking.   And at last we were in the
snow-packed, tree-strewn river gorge.   Following it
downwards to where the snow terminated, and the
long imprisoned glacier torrent foamed out, muddy
and furious, to follow its noisy course down the valley,
we ascended the steep bank to the left and were soon
at our camp.   Fortunately before leaving camp in the
morning we had put some firewood into the tent;
this was now quite dry, though all the rest was
dripping ; with its aid we quickly got a fire going,
and splitting other logs so as to expose the dry inside,
we began to steam before a roaring fire.   Our
shirts dried quickly under this treatment, and as our
coats were dry and the rain had ceased, we cooked
our bacon in comfort, and fried a scone in the remaining fat in the frying-pan, and after our day of

twelve hours' work felt perfectly ready to turn in by
the flickering light of the dying fire. One or two
corners of stones in my bed seemed determined to
make a lasting impression on me; chipnuncks began
their nightly scrambles up and down the outside of
the tent; I had some dread that a mosquito or two had
eluded our vigilance and got inside our defences of
netting; but all these troubles quickly vanished into
the blissful atmosphere of dreamland.

CHAPTER XI.

" The joy of life in steepness overcome
And victories of ascent, and looking down
On all that had looked down on us."

TENNYSON.

An early start.—A steep couloir.—A *mauvais pas*.—The summit of Mount Bonney.—Sunset.—Benighted in the forest.

IT seemed now quite evident to us that Mount Bonney was not to be conquered without a really big effort, so to prepare ourselves for that, we determined to make the 8th an off day. It was also necessary to fetch up more provisions from our *cache* down at the mouth of the valley. H. volunteered to do this, and I remained at camp to work up notes and sketches. To be alone in this wilderness of forest and cliff, glaciers and mountain torrents, bright wild flowers, bright sunshine, and the weird cry of the marmots, and with leisure to let the mind dwell on it undisturbed, was an experience well worth a day, even if no other reason for pause existed. After breakfast

H. started down the valley, and returned in the evening with some meal and meat-tins and his half-plate camera. While he was away I was able to shoot a marmot and a little chief hare, and had them stewing for supper when he arrived. A spring of clear water, in the midst of sweet-scented, large red-flowered mimulus, oozed from the rocks near our tent, and fetching water thence we made the kettle boil on the cedar logs.

After supper we took a last anxious glance at Mount Bonney, rising from its bed of glaciers in dark cliffs to a height of 6,500 feet above our camp. Cold, grey wreaths of fleecy clouds wound in and out through its gullies, illumined here and there by shafts of lurid sunset light. The weather and its promise for the morrow filled our minds with many forebodings. I felt certain we were in for wind and rain, so I did not change the plates in my camera. If it should turn out fine H. had six plates all ready in his. With what hopes we could conjure up we soothed our minds to sleep, and soon the roar of the torrent and all other sounds were as though they were not.

I woke once or twice and looked at my watch. At 3.30 it was time to get up. Anxiously we stepped out of our warm bags into the chill morning air. Stars were twinkling brightly. Mount Bonney looked dark and sullen, but its *aréte* was clear cut against the sky; lower down, grey mists lay like a blanket on the glacier. The flowers and trees were dripping with dew. All this looked promising, so shivering in the cold we lit the fire, made a cup of tea, and shouldered our swags which we had prepared the night before, closed the tent doors, and shortly after four o'clock started on our way. H. carried his half-plate camera and the rope; I took the provisions and the plane table. On former expeditions we had found that in making elaborate plans to avoid difficulties, we had involved ourselves in worse ones which were unforeseen; so now, as we intended to make our attack by the glacier descending from the col to the north-westward of our peak, we determined to take a "bee line" for its lower termination, and let difficulties come when they might. After a small strip of forest, our way lay across an open area of large angular boulders; then came a very tangled piece of forest, which we had to follow down into a ravine and then work up a steep ascent beyond. Our aim was to strike the stream from the glacier as soon as possible. Leaving the tall forest, a desperate struggle ensued with alder scrub through which we scrambled for an hour. There was evidently a bear in it, for we saw the branches swaying to and fro as he pushed his way along a little below us. At last we entered the hollow worn by the glacier stream; it was all choked with compacted snow, which gave us good travelling towards the glaciers. Immediately on our left, above a slope of

débris, rose steep cliffs, forest-clad above, and in their
lower portions pierced by several dark caves. Through
the grass and wild flowers on the shingle slope, well-
marked paths converged to the entrance of these caves;
they were evidently the home of bears. We had no
rifle, so we gave them a wide berth, for to meet a
female grizzly with cubs was an adventure we felt
disinclined to tackle with nothing but ice-axes. Louder
and louder rose the roar of a waterfall, and turning a
bend, we came on the open torrent making a fine
leap of a hundred feet, from the flowery slopes above
into a tunnel it had bored for itself beneath the snow.
We had now to begin a climb through tangled forest
once more, but reaching the level, about 200 feet
higher up, found a fine open, clothed in a perfect
meadow of *Veratrum viride* as high as our waists. Here
again a bear was making tracks ahead of us. The broad
leaves were so filled with dew that walking through
was as wetting as if we were in the river, we therefore gladly followed the path he made; the stems being
crushed down and the dew shaken from the leaves in
his wake. For several hundred yards he had gone
just in the direction desired, but when he turned off
to the slopes on our left we had to say good-bye to
him, and take the shortest line to the shingle beds
near the stream, which stretched to the foot of the
glacier. The cliffs on our left continued to rise higher
and higher. Now the forest cap above had given

place to one of glacier ice, from which seven fine
waterfalls leaped down amongst the vegetation below,
and made their way by countless channels to the
larger stream. The whole air trembled with the roar
and splash of torrents and cascades. As yet there
was enough forest in the scene to give richness to it,
so we found ourselves in a perfect Alpine paradise,
which no being higher than a bear had ever entered
before.

On reaching the recent moraine near the foot of
the glacier, and noticing how much the glacier had
shrunk in recent years, we halted for some refreshment. It was now four hours since we had left the
camp, and we had risen just 1,600 feet. Following
the side of the glacier as far as possible, we were soon
forced by the closing in of the cliffs to take to the
ice. This we found so steep that without endless
step-cutting we could not ascend by it; we therefore
crossed it to the northward, and followed the moraine
under the cliffs of the Ross peak range.

The strata in these cliffs is nearly vertical, but
contorted and arched in some places in wonderful
style. Caves again were numerous at the base of
these cliffs, and looked like the habitations of bears.
Once more the terminating of the moraine compelled
us to take to the glacier, now covered with snow, and
at 9.30 we found ourselves cutting steps at the foot
of the couloir leading to the pass. The steep slope

being scored by tracks of falling stones from the cliffs on either hand, we commenced our ascent in the centre.

After cutting a few steps and zigzagging upward, we looked up at the cornice overhanging the top; it seemed a long way off, and our hearts, I fear, failed us a little. I had hoped we might have got grips for our feet. To cut steps all the way up was, for us, next to impossible. Then I found we could get a little grip by kicking, and soon the snow became softer and our toes went well in. Holding good grips with our axes we now ascended at a fair pace. A bergschrund near the top forced us once more on to the rocks; we thus avoided the cornice, and at 10.30 just two hours from the foot of the glacier, we stood on the col. Before us the mountain fell away in precipitous slopes to a glacier-filled valley whence streams flowed into the Illecellewaet; and we could detect the railway track as a fine line, far below in the latter valley.

The two sides of the col presented a marvellous contrast; on the north a heavy cornice overhung a snow-filled couloir leading down to a glacier. On the south a perfect garden of Alpine flowers was in full bloom. There was the familiar *Dryas octapetula* and gay yellow *Haplopapus Brandigrii* all low-growing plants. Deep down however beneath the flowery slopes the valley was filled by a small glacier. The *arête* on

our left leading towards Mount Bonney was broad
and free from snow, and without any delay we resumed
our ascent along it. The slope was gentle, and for
half an hour nothing in the world could be easier.
We could not see very far ahead owing to a series of
knobs, one of which always rose a short distance
ahead. I knew it was not all going to be easy like
this, so we hurried anxiously upward, hoping soon to
come in sight of the little curved peak visible from
the railway, and which we feared would prove a
serious obstacle to our progress.

Scrambling up some angular sharp-edged blocks
of quartzite into which the *arête* had now contracted,
the curved peak came into view, and the look of
it was by no means reassuring. From where we
stood, to its foot, the *arête* was very sharp, flanked
on the southward by steep snow slopes leading down
for about 2,000 feet to the glacier on our right, while a
heavy cornice overhung the almost vertical precipice to
the northward. From this *arête* the peak sprang upwards in nearly perpendicular crags, snow-seamed, for
over 200 feet. Two possible routes were all that offered :
one was, to scale the apparently vertical face in front ;
the other was to skirt round the peak on the steep
snow slope to the right, and so turn its flank.

The slope was so very steep and the snow so likely
to slide, that we decided the latter route would be too
risky, we therefore put on the rope and pulled ourselves

together for a stiff climb. Having rested for a few minutes and deposited our spare food under a boulder, we started along the cornice with much caution, and then began to climb upwards. There was just enough loose powdery snow on the crags, to make it most difficult to find a firm grip for either hands, feet, or axe.

The projecting shales, set vertically, were also so rotten that at every step we had to dislodge quantities of rock ere we could find any solid foothold. Every move needed the greatest possible caution, for we could not avoid being in a direct line one over the other. The ridge about half way up, divided into two parallel ridges, the right hand one composed of bare crags, completely overhanging the snow slopes below, while the other was more or less a continuous snow *arête* to the summit.

After much scraping away of snow with my axe I succeeded in reaching these crags while H. continued his way up the snow *arête*. We kept the rope tight between us and ascended abreast, he holding on while I sought out fresh grips, and when he moved I made myself as secure as I could. The crags on my ridge soon became more trustworthy, and at ten minutes to one o'clock the top of this first peak was beneath my feet. A snow cornice overhung the *arête* H. was on, and as he sung out from below that he had no grip whatever in the loose snow, I gave him a good pull, and up he came making a fine gap in the cornice. We were up now, so much was certain, but the glance which passed from one to the other expressed the foremost thought in our minds, "What about the getting down!"

The view from the curved peak was superb. A perfect ocean of peaks and glaciers all cleft by valleys, and the main peak of Mount Bonney still rising in a dome of snow to the eastward. The weather looked threatening. Most of the landscape was bathed in sunshine, but there were heavy clouds hanging about the peaks, and one drifting towards us looked so lowering that we feared a thunderstorm. Our first thought was to hurry up with the camera, but ere we could get it fixed the clouds broke in a furious shower of hail accompanied by strong wind, and the photograph taken under such circumstances was decidedly of a shaky appearance. The gap in the cornice through which H. had ascended was distinct enough, but the distant view was all doubled and confused.

As quickly as it came the storm passed away, and descending an easy slope of snow for a few hundred yards we commenced the ascent of the final peak. It was now nothing more than a tiresome trudge up steep domes of snow. When one was reached which we hoped was the final one another loomed up ahead. At last the highest crest was in sight, with a huge

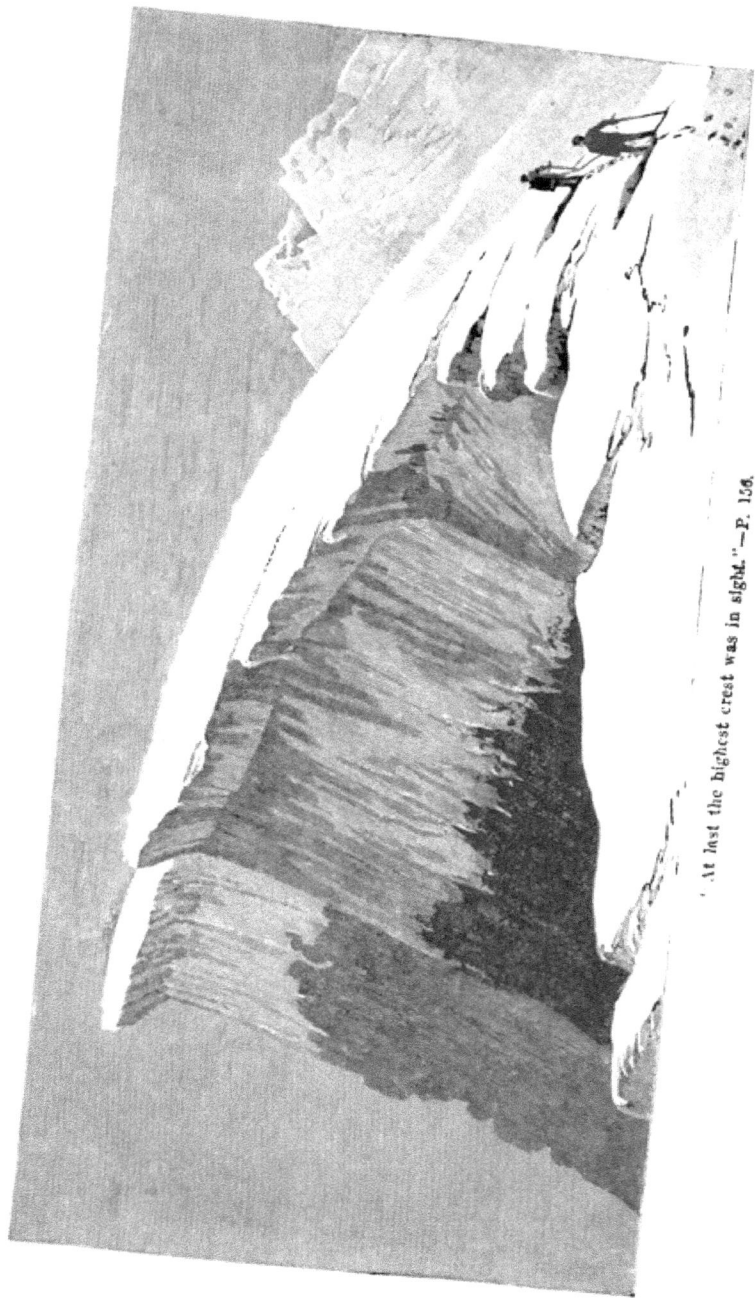

"At last the highest crest was in sight."—P. 156.

cornice overhanging the cliffs. Inside the cornice a
narrow ridge of crags made themselves visible through
the snow, and at ten minutes past three we were on
the summit.

We placed the thermometer in a suitable position
and took the reading of the barometer; it showed us to
be about 10,600 feet above the sea, probably a little
more than this, and over 6,000 feet above our camp.
It was too late in the day to admit of unnecessary
delay, so I set up the plane table, took a series of
observations, and then we turned our attention to
photography. As much of our view, in the direction
of what was most familiar in the panorama, was shut
out by the cornice projecting from the summit,
H. ventured out on it, while I, taking a round turn of
the rope on a crag held him firm. With his axe
he pushed down some of the cornice, and fixing the
camera took a photograph of the peaks of Mount
Sir Donald and the *névé* of the great Illecellewaet
glacier. The horizon line formed by the range of
the Rocky Mountains was also visible in the photograph. We took other views with less difficulty.
The unknown valley with its river[1] glistening
like a silver thread, was in view for full thirty

---

[1] Mr. Bailie-Groman, writing in the *Field* for May the 11th, 1889,
comments on our map, and considers that this river whose glacier
sources we surveyed is the Lardo (or Lardeaux) which flows into the
Kootenay Lake. His long experience of the Kootenay country renders
it likely that he is right, so I have entered this name in my map.

miles. To take in all the points of the panorama would have needed hours of work. Unfortunately we dared not spend this time; the sun was now going down to the westward, and it seemed already beyond hope that we could regain our camp before nightfall. The rocks of the crest were composed of a most beautiful, fine-grained, white quartzite, speckled with a few dark spots of oxide of iron or manganese. The joints were sharply rectangular, giving an almost artificial appearance to each block, and as we stirred them in building a very small cairn they rang with a metallic sound. At 4 P.M. we commenced the descent, and going as fast as possible, between glissading and running we were soon down to the col, beyond which the curved peak rose to the westward. As the evening sun was now shining on the side of the peak by which we had ascended, we felt that, soft as the snow had been in the morning, now it would be all slush, and the bad bit consequently much worse than before. We thought anything would be better than to attempt such a descent, so we determined to try and turn the peak in flank and cross the steep slopes of snow, plastered on to its face, which we had carefully considered during our ascent.

Accordingly, we bore away to the left, descending to a shoulder of the ridge below the peak. On reaching it we found ourselves on the brink of the precipice overlooking the glacier-filled valley

to the westward, and it too was topped by a cornice.
Farther to the right the *névé* we were on curved
downwards, and though nearly vertical in its face,
there was no actual cornice. It looked an exceedingly uncomfortable bit of work, but our only
choice lay between it and what seemed the worse
descent over the summit of the peak. The question
was, could we reach the snow slope below the brink of
the precipice? and having reached it, would it bear
our weight? H. buried himself as deeply as possible in
the snow, and when he considered himself quite firm
I turned my face to the slope, and holding on to the
rope kicked my toes in and went over the brink. I
took the precaution, too, of burying my axe up to its
head at every step. Just below the brink there was a
projecting crag. This I thought would give a firm footing before testing the snow slope. I got one foot on to
it, and was taking it as gently as possible when the rock
gave way, a large piece of snow went with it, and fell on
the slope twenty feet below. I stuck my knees into the
snow, but felt my whole weight was on the rope. Then
I heard a swishing noise in the air, and glancing downwards saw that the whole snow slope had cracked across,
and was starting away down towards the valley in one
huge avalanche. H. hauled cautiously but firmly on
the rope, and getting what grip I could with toes,
knees, and ice-axe I was quickly in a safe position,
and the two of us standing side by side, watched the

clouds of snow filling the abyss below, and the huge masses bounding outwards. We listened to the sullen roar which gradually subsided, and all again seemed quiet except that a few blocks of consolidated snow went careering along, down the glacier, for some time, after the great mass of the avalanche had come to rest. This route was manifestly impracticable. There was now no choice. We must retrace our steps to the summit of the curved peak, and go down by the same road that we had come up. We had eaten nothing since a few mouthfuls at 11 A.M., so between anxiety as to what lay before us and hunger, we felt far from happy. Never did anything feel more weary than that plod up the snow slopes to the peak. There we sat down to rest; I searched my pockets and found a small packet of tea and one cigarette. H. ate the tea, and I enjoyed the cigarette, and feeling our nerves in a more reliable condition we commenced the descent.

As far as it was practicable we went down by the crags avoiding the snow, and made each step as secure as possible by shoving tons of loose slates and shales over the precipice. Then we had to quit the rocky ridge and cross the little snow-filled couloir to the other ridge. The snow on this was the chief danger, for it would not bear the slightest weight and it covered up the sharp loose slates. The axes were no use to us, so taking off the rope we tied them together and lowered them down, then making a bowline hitch on

" H. hauled cautiously but firmly on the rope."—P. 159.

the other end of the rope we hung it on to a crag, and
with its help scrambled down fifty feet to another firm
foothold.   A smart chuck brought the end of the rope
free, and hitching it on again, we reached with its
help the more secure portion of the ridge, and felt
once more happy for all danger was past.

It took us some minutes to reach the place where
we had deposited a few biscuits and a little beef in a
tin, and then hurrying on we regained the summit
of the col at 6.30 P.M.   As we crept down the rocks
towards the snow-filled couloir, we could not resist
pausing to admire the marvellously beautiful sunset
glow, which had flushed the whole range of the Rockies
with bright carmine ; while the nearer peaks and
glaciers glowed with deep crimson.   Never before or
since have I seen such intense evening tints.   Night,
however, was close at hand, so on reaching the snow
we glissaded, and ran downwards. Crevasses then forced
us on to the moraine.   Instead of following our course of
the morning we determined to keep to the left side of the
glacier and torrent, and take chance for the difficulties,
but it was obviously shorter.   Running and leaping
from boulder to boulder, wading streams, taking a
straight line through everything, and making many a
stumble amongst falling stones, we found ourselves at
last, with much-bruised shins, but fortunately without a
sprained ankle or broken limb, at the margin of the
forest.   It was now twilight, and this side of the valley

M

was unknown to us, so as closely as possible we
followed the course of the streams. When it plunged
down in a waterfall, we slung ourselves downwards
through fern and alder bushes by its side.

Ere we reached the ravine where the river was
arched over with snow, night was upon us; but we
had fixed a pole in the snow at the point where we
left the forest, as a guide for us where to enter it
on our return. The sky was overcast, and so dark was
it now, that only by groping along did we find the
pole; and, leaving the river bed, we entered the alder
scrub in pitchy darkness. It had been bad enough in
the light, but now in the dark it was simply heart-
breaking. Never could one be sure of a footing on the
slippery stems, and a fall every now and then nearly
shook the life out of us. We hoped that we might not
tread, by accident, on the tail of a grizzly, but took
comfort at the thought of their deficiency in such
appendages. We steered our course by the sound of
the torrent, and by looking backwards at a certain
peak which showed clear against a patch of sky. Then
we were in the high pine forest, feeling with our axes
for fallen logs, and fending off branches from our eyes.

Once or twice we almost despaired of getting through,
and thought of sitting on a log until morning. We
could see nothing whatever, in fact I kept my eyes shut
most of the time, and only now and then glanced
over my shoulder to see was the sky visible and the

peak we were steering by. In the high woods there was soon no use in looking out for the latter, so we steered solely by the sound of the torrent. At last a white line was recognizable in the valley ahead, which we knew was a heap of boulders near our camp, and

> " Be the day weary
> Or be the day long,
> At length it ringeth to evensong."

11 P.M. found us round a blazing fire, sipping chocolate and picking the bones of a marmot. And so our long and successful day came to its close.

CHAPTER XII.

"For, the man—
Who, in this spirit, communes with the forms
Of nature . . . . he cannot choose
But seek for objects of a kindred love
In fellow-natures and a kindred joy."
                                        WORDSWORTH.

Fetch camp from Loop Valley.—A breakfast party.—Illecellewaet.—
A miner's camp.

AFTER a good night's rest on our bed, in which the angles of the boulders made themselves very distinctly felt, we rose refreshed, and after breakfast made up two light packs and set off down the valley. As the stream was not at this early hour swollen to its full height, we were able to follow its bed for most of the way, and so avoided the difficulties of the forest. We thus reached the bottom of the valley in about an hour and a half from our camp, and, ascending the slope at the side of the high trestle-bridge, gained the railway track, which we followed till we reached Glacier House. A good dinner came in well after

camp meals, and oh! for the delight of getting out of
one's clothes after being in them for nearly a week,
and the luxury of a bath, to say nothing of a bed
without boulders for feathers.

Next morning we set off down the track again to
fetch our tent and blankets from the Loop valley. A
number of men were engaged on this portion of the
railway, felling timber, building a snow-shed and
repairing the line. While H. returned to Glacier to
fetch a part of his camera which he had left behind, I
accepted an invitation from the workmen to join them
at breakfast. Amongst them were the two youths who
had been our companions on the glacier field. Others
of them had long experience in the mountains, and their
yarns were full of interest. One rather elderly man
said that he had just returned from a fine spree at
Donald. He had got invalided with bronchitis and
was sent down on full pay to hospital; but on reaching
Donald he found that there was a bad fever case in
the hospital, so not caring for such company he met a
"pal" with whom he went on the spree; they spent all
their money, and now he was back to his work feeling
quite well. It did not seem a good argument for teeto-
talism, and was possibly a somewhat inaccurate account;
but I mildly suggested that he was an old idiot, at his
time of life, not to be laying by some money for a rainy
day; and, as this was Saturday, I asked him to come up
and join us at divine service the next afternoon. He

promised to do so and "would also bring some of the
boys along." When Sunday came he was as good as
his word.

As we went on our way down the track we met
another gang of fine-looking fellows coming up to work
at felling trees. Two days afterwards they were swinging
round a tree to shoot it down the mountain side. It
slipped before they were ready, five of them were
knocked over, more or less hurt, and one had his
brains dashed out. Such is life!—in the Selkirks!
As we were both armed with our cameras we took
a series of views on our way up to camp; and then
halting to enjoy our last meal in the Loop valley
we shouldered the remaining swags and reached Glacier
House for supper.

We had heard much of the mines in the mountains
near Illecellewaet, about fifteen miles to the westward,
and as Mr. Corbin, the owner of some of them, invited
us to go down to him for a couple of days, we determined
to accept his invitation. He promised us horses to ride
over a high range whence we could get good views
of the mountains we had been at work amongst, and
which would help us much with our map. Sunday was
spent at Glacier House, and on Monday, August 13th, we
left for Illecellewaet in company with Mr. Bell-Smith
who hoped to make studies for future pictures. Mr.
Corbin met us on the platform and we spent a pleasant
evening in his shanty with his friends, all of them

concerned in prospecting and mining. Of all occupations, that of the prospector seems to be the most
attractive. Once a man tastes the charm of the wild
woods and secluded mountain glens, with the adventures incident on such a life, and has all this stimulated
by the prospects of a trump card turning up in the
shape of a lode of silver ore or a sand bar of gold,
everything else in the world seems flat. It is wrong to
think that such men are always dare-devil desperadoes,
given up to wild dissipation and excess, like the picture
of them usually drawn in story-books. Here we
were in the midst of prospectors of the most enthu-
siastic type, all ready to face cheerily any hardship
or danger, and as good fellows as you could meet with
anywhere in the world. Nature in its wildness had
humanized them. Its beauty was the charm of their
lives, and the language by which every tree, and plant,
and rock, and torrent spoke to them, had become so
much a part of their existence, that life on the plain
or in the centres of civilization would be for them
the same as banishment.

As Mr. Corbin had no shake-down for us, we went
to sleep at one of the inns. In fact the only one
running; for in consequence of bad work going on,
the sheriff had come along and shut up the other.
The washing-basin for all guests was in the yard
on a tree root. A strip of leather nailed to the side
of the house and a piece of broken looking-glass

provided shavers with all the necessaries of life, but
we were snug enough, though the house stood on
legs in a big pool of water caused by the overflow
of the river. Some trees had been felled to make
room for the "city," but, as the custom in British
Columbia is, when felling a big tree to fix in its side
a spring board on which the men stand, and cut it
across at about eight or ten feet from the ground, they
leave a stump most difficult to get rid of, and here were
these great stumps, charred by fire mixed up among
the wooden houses. The city seemed to have no
plan, but on the map which we saw it was laid out in
the most splendid series of lots, and two steamers
were represented as plying on the river—which, by
the way, is a glacier torrent flowing at about twenty
miles per hour.

Late in the evening a train of thirty mules arrived
from the mines belonging to the Selkirk Mining
Company with heavy loads of galena fixed on the
pack-saddles by means of the famous diamond hitch.
The packs were taken off, laid in a row before the office,
and the mules turned into an inclosure, where they
rolled and refreshed themselves after their long day's
work.

Next morning after an early breakfast Mr. Corbin
had four horses ready, one for each of us and one
for himself. His companion Mr. Ben Macord, known
as "Mountaineer Ben," accompanied us on foot. Macord

had devoted his whole life to prospecting and its kindred occupations. Every part of British Columbia was known to him. As a boy he had served on the United States Boundary Commission. In Riel's rebellion he had served on the prairie, in the Rocky Mountain Rangers; and after a little conversation we were fortunate in being able to persuade him to come back with us to "Glacier" and help us till we left the mountains.

Our route lay up a zigzag path made by Mr. Corbin through the forest, on the steep northern slope of the valley.

We rose steadily, gaining more and more extended views of the valley, and when past the forest we entered upon grassy Alps, ablaze with scarlet and yellow flowers. Such an Alpine garden I have seldom seen.

The path now and then was across the face of slopes so precipitous that our guide advised us all to dismount and lead the horses, which we did; and at length when the aneroid registered an elevation of 4,000 feet above the railway the summit of the pass was gained. From this point our view extended up the valley of the Illecellewaet to Mount Sir Donald. Mount Bonney and the Dawson range presented an entirely new aspect, and the mountains and lateral valleys on the other side of the Illecellewaet formed a fine panorama.

Besides this view over ranges now familiar, there was

that on the other side of the pass, of which up till now
we had seen nothing.   Here we looked down upon the
north fork of the Illecellewaet with its numerous glacier-
fed tributaries, beyond which some snow-seamed rock
peaks rose into strange pinnacles and crags.   The sides
of these valleys were clad in an almost unbroken
covering of forest.

I delayed to take a few observations and photographs,
and Mr. Bell-Smith began a picture, but Mr. Corbin
hurried us off, for he had sent word to have dinner
ready at the mines and feared it would be spoiled.
We might, he said, have what time we liked here on
our way back.

The path now went down in zigzags through a narrow
glen, and when about 2,000 feet from the summit, we
reached the mines.   They consisted of a few short levels
driven into the hill side, with heaps of quartz full of galena
piled near the entrance.   Seven men worked these levels,
and all lived with the "boss," his wife and child, in a
tent built on a little platform of logs, forming a kind
of bracket on the steep mountain side.   Our hostess
was from Dowlais, South Wales; she had been eighteen
months in America, and twelve of it had been spent in
this lonely valley, never before entered by a woman.
The youngster had not much room for a playground,
as the whole place was one continuous precipice.   The
dinner she gave us and the hot cake and stewed prunes
were most excellent.   In 1884 the first prospectors of

these valleys (Mr. Corbin was one of them) encountered great difficulties, privations, and misfortunes. They were but fulfilling their destiny in finding new abodes for mankind; and here was the first woman following up their work, and turning a mere resting-place into a comfortable home. The next move towards civilization and reclamation of the wilderness, will be the making of a road up the valley. Until that is done the produce of these rich lodes of ore will be next to worthless.

As we were anxious to have some spare time on the pass, we made but little delay at the camp. At starting we nearly met with an accident, for the horse Mr. Bell-Smith rode was accustomed to go first, and on being placed further back in the line he objected, and attempted to rush past the others where the path was so narrow that to pass was impossible. One horse was of course swung round and his hind quarters shoved off the path. He held on well with his fore legs, but only after a desperate scramble succeeded in regaining the trail. The horse accustomed to lead was now placed first, and then they all toiled along patiently and steadily up the zigzag path. A " fool hen" got up from some scrub and alighted on a low branch of a pine. I took a photograph of him, and then tried to knock him down with a stone. He was about four yards over my head, and a large stone I threw hit him and pushed him off the branch, but he clung on by his claws and fluttered back into his former position, and went on cran-

ing his neck from side to side in the most inquisitive
manner, exactly as though he were asking "What on
earth did you do that for?"   I merely flung a handful
of earth at him in answer, and away he flew, crowing, to
the other side of the ravine. Near the miners' camp I had
seen a tiny humming-bird flitting from flower to flower.
These were the only birds seen by us during our day's
ride. On reaching the pass, Mr. Bell-Smith sat himself down to paint a picture, H. to take photographs,
while Mr. Corbin and I strolled away up the mountain
to the eastward, following a path worn by the feet of
mountain-goats and cariboo.   Mr. Corbin had many
a tale to tell of adventures in the wild valleys beneath
us, and of the dangers encountered while scrambling
about these huge precipices, searching for lodes of
ore.

These game paths, such as we were following,
continued along all the mountain ridges, and often gave
prospectors the clue by which they found their way
out of all kinds of dangerous places in safety. On
many of the eminences we gained, and wherever there
was a little plateau commanding a view of the ravines
below, there was always a bare patch which Mr. Corbin
said was the stamping place of the wild goats;—here
the sentinel of a little flock will stand looking out for
danger and stamp with excitement till the ground is
beaten perfectly hard and bare. On this ridge I met
with, for the first time, a beautiful little Alpine flower,

Erigonum umbellatum, it was here growing in great
quantities. Having made a sketch and taken another
photograph we rejoined our companions, and then
resumed our journey, descending over the flowery Alps
to the dark pine wood, and then on down to the railway
through bush and forest where the path was cumbered
by many a fallen log which the horses had to jump in
succession.

Another night was pleasantly spent in Mr. Corbin's
shanty. Many pipes were smoked and many yarns
spun. Amongst those who dropped in to chat was a
certain " doctor " from Colorado, connected with some
smelting firm there, and he had come up to see whether
the output in the Selkirks would make it worth
while starting a branch of their business here. We
were asked many questions as to the big valley we
had entered to the southward, and we discovered that
it had that very week been entered by prospectors from Illecellewaet and that they had called it
Fish Creek. They had entered this big valley by
ascending Flat Creek valley, and struck it much farther
from its head than we had.

Next morning we found that there was a special
train coming up the line from the westward, with
Government inspectors and that we could get a passage
in it. So we hurried over our breakfast to be ready.
Ben Macord was there too—ready, just as he stood
in his shirt-sleeves and trousers, with no luggage

but his miner's pick; whether it was for a day or a
week or a month, it was all one to him. Soon the train
arrived with the inspectors sitting on a truck covered
by an awning, pushed along in front the engine. We
took seats in the caboose, but our journey was very
tedious, for we stopped at every trestle-bridge and they
were innumerable. Each had to be inspected and
measured, and then on we went. And so at last we
reached Glacier. In this trip we had not only gained
valuable observations for our map, but we had got a
glimpse of human life, full of interest to us, different
from anything in the old world, and sure to return often
to our minds as one of the pleasant memories of the
Selkirks.

CHAPTER XIII.

" This is the forest primeval. The murmuring pines and the hem-
        locks,
    Bearded with moss and in garments green, indistinct in the twilight,
    Stand like Druids of eld, with voices sad and prophetic.
    Ye who believe in affection that hopes, and endures, and is patient ;
    Ye who believe in the beauty and strength of woman's devotion,
    List to the mournful tradition still sung by the pines of the forest."
                                                        LONGFELLOW.

Start with horse for Beaver Creek.—Camp on the Tote Road.—Ford
        Bear Creek.—Impenetrable forest.

Two expeditions now presented themselves to us as the
most important for the prosecuting of our pioneering
work. One was a journey from Glacier to the upper
portion of Beaver Creek, on which we hoped to have
opportunities of mapping out all that face of the main
range which we had not as yet seen ; of inspecting the
eastern side of Mount Sir Donald ; and learning more
concerning the curious plateaus which we had seen from
the *arête* of Sir Donald, and which we heard were known
to hunters as the Prairie Hills. The other expedition
was one concerning which Ben Macord was very keen,

from a prospecting point of view, the exploration of the valley beyond the Asulkan pass. On both these expeditions we hoped to use the pack-horse and, as Mr. Marpole had kindly sent up men to clear the trail in the Asulkan valley, and they were now at work, we decided on taking the Beaver Creek expedition first.

On August 16th we made up packs with a tent, blankets and provisions for three ; packed the cameras and surveying instruments in our knapsacks, and placing with these a rifle and our axes we were soon ready to start. About nine miles from "Glacier," to the eastward, was a station called Bear Creek, and, as this would be our starting point from the line of railway, it was decided that H. and Ben should walk on with the horse in the forenoon, and that I should wait for the East-bound train in the afternoon and take on the packs to Bear Creek.

To lead a horse from "Glacier" to Bear Creek was no easy undertaking ; at a hundred yards from the inn the difficulties commenced, in the passage of the glacier torrent. Our first idea was to ford this stream, and we went up and down the bank to find some suitable place. We should never have dreamed it possible for a horse to get across alive, but Ben had such large experience of this kind of work that, in all these matters, we trusted to him and usually found he was right. This time, however, it was no go. The horse would not take to the water; and as he, too, had had large experiences of mountaineering in the Rockies, we placed much confidence in his opinion, and gave up the idea of fording as impossible.

The only other thing for us now to do, was to take the horse up the embankment and cross the high trestle-bridge, on the railway. Two difficulties again lay in our road; one was that the bridge was not floored, it was merely a series of transverse bars, and the second was that while we were on it some stray locomotive might come along and knock us all to bits. To guard against this latter danger we got the agent at Glacier to telephone along the track in both directions to know if the course was clear. Then getting two pieces of plank we laid one down on the bridge and induced the horse to walk on to it: when he stepped on to the next board, we took up the first and laid it down in front, and when he moved on to that we shifted on the other, and so slowly, after shifting the boards about ten times, our cayeuse was safe on the further shore.

Leaving H. and Ben to go on their way and cross all other streams and bridges as best they could, I returned to the inn, and when the train arrived about 2 P.M. I took the packs on board, and crossing Rogers pass left the cars at Bear Creek. Several times I looked out from the train for the horse, but in vain; and it was not till I had waited for an hour and thought of all kinds of accidents that might have befallen them,

N

that my companions turned up, with a volunteer helper who had joined them *en route*.

As the evening was advancing we immediately set to work to pack the horse, and as Ben was an adept at the diamond hitch, the packs were put on so that they could neither loosen nor fall off. The mysteries of this famous hitch are too complicated to explain here, and, as it was not till we had done the thing about ten times that we began to do it right, its complexity may be judged.

Roughly speaking, the method of packing a horse right is this : you make up your goods into three packs, one goes on each side of the pack-saddle and one between them, on top. The weight of the side packs must be equal; that is the first important point. The packs are attached to the saddle by slight cord, and then comes the most important thing in the whole affair. This is the synch, or belly-band, to one end of which is spliced a long rope, and at the other end is a hook through which the rope, after passing over the packs, is rove, a series of bights almost like the game of " cat's cradle," surrounds all the packs, " fore and aft," and then by putting your knee against the horse and hauling on the last end, everything jambs itself tight, and a single hitch of the end to the synch makes everything secure. The rope we used was one which had seen some vicissitudes of fortune in my company. Its first good work was to save the lives of some of

our party in a bad slip, near the summit of the Balm-
horn in the Bernese Oberland. It was next used as
the mizen topping-lift of a fifteen-ton yawl. It was
my tent-rope in the New Zealand Alps. It was the
bridle used on a deep-sea trawl that went down to
1,000 fathoms beneath the surface of the Atlantic.
It trained a colt. Now it was in our diamond hitch,
and I regret to say that its old age was disgraced,
by its being used for cording one of my boxes on the
voyage home.

At Bear Creek station the railway runs high above the
forest-filled valley. And as our first desire was to reach
the bottom of the valley and cross the river, 1,000 feet
below, we had to look for some practicable route for the
horse. We were told that the "tote road," as the
track was called, which was made for the purpose
of bringing up material for the construction of the
railway, was low down in the valley and in good repair.
Could we but get down to this we might follow it to
where the river might be fordable.

Our new-found ally said he knew the way down.
We soon proved that he did not. Then we went along
the railway to the eastward, where a construction train
was on a siding with a gang of men living in it. We
interviewed the "boss," an Italian, and he told us to
follow on as we were going, and when we came to a
certain big boulder, to strike into the forest on our
right, and we should soon hit a descending trail. There

N 2

were a good many big boulders, so when we recognized
one as specially "big" we struck into the forest and
found ourselves in a perfect maze of fallen trees. The
forest, too, was half burnt out and still smouldering, so
between dense smoke, black dust, fallen logs, and
sharp broken branches, it was just the place *not* to
have a horse in. However Ben faced everything.
He led the cayeuse up to logs which he leaped
most skilfully. Then crash, crash! down the steep slope
through charred branches we made our way. The
trees every now and then became quite insurmountable,
and we struck off now right, now left. Then we were
completely stopped, and we had to face back up hill
again. The difficulties at last seemed so great, and the
descent, so precipitous, that lest the horse should get a
bad fall, we halted. Ben and I stayed with the horse,
while H. and the volunteer went off to explore for a
trail. As long as possible we kept up communication
with them by shouting, but they got beyond hearing, and
after waiting half-an-hour and hearing nothing of them
we started again, going straight down the mountain
side. The horse got some nasty falls, but Ben always
managed to keep a grip of his halter. At last we saw
a trail below us, simultaneously we heard the shouts
of our companions; they had evidently struck it further up, and were following it down, and after a
few leaps, slips and tumbles amongst the fallen
trees, we reached the trail with loss of nothing save

the skin of our shins and sundry bits of cloth from
our garments.

The trail we were now on led us down to the
tote road, which, though covered with weeds and encumbered a little by fallen trees, promised us good
travelling in the direction we wished to go. Along the
sides of the road raspberries grew in profusion, and
the delicious fruit was most refreshing after the heat
and toil of the descent. As darkness was closing in
we could delay little, but following up the road to the
westward we reached a stream, near some ruined log
huts, and there in the middle of the road we pitched
our camp, lit our fire, and said farewell to our new
acquaintance, who struck straight up through the
forest towards the railway track. We boiled some
rice and ate it with squashed blueberries, raspberries,
and sugar, and so made a most luxurious supper. And
then went to sleep to the music of the waters of Bear
Creek.

The early morning was, we knew, the best time to ford
the stream; for after the snow slopes had been refriger-
ated by the cold night air, and the supply of water thus
lessened, the creek would be at its lowest. We had
not as yet seen the river, though its roar filled the air.
After an early breakfast we broke camp at 6.15 A.M.,
and leaving the tote road struck down through the
damp, rank undergrowth of the forest. Here the
devil's club grew luxuriantly, and was gay with its

tufts of coral red berries. The descent into the gorge worn by the torrent was most difficult, but eventually we found a safe route for the horse, and reached a shingle flat strewn with fallen trees, one of which completely bridged the torrent, which was foaming wildly over its stony bed. Selecting a spot where it seemed that the horse could cross most easily, we took his packs off, and hitching the halter on his neck headed him for the stream. He went on step by step with the greatest caution, until reaching a shingle flat in mid-stream he refused to go any further. We expended all our blandishments on him in vain, even the stones we threw had no effect. Ben suggested my crossing over on the fallen tree, which no sooner had I done than the horse entered the worst current in the whole passage, and though nearly carried off his feet and swept to destruction, he reached the shore in safety and came to me. The whole cause of his halting in mid-stream was simply that he saw there was no one on the other side to receive him. We found no little difficulty in balancing ourselves on the fallen trunk when crossing; but Ben, more used to such work, walked over with apparent ease, carrying on his shoulder one of the horse's packs, and then went back for another.

The blueberries growing here were the largest and richest we had yet come across, but we dared not spend time gathering them with all kinds of

unknown difficulties ahead. We were now at the very
bottom of the ravine formed by the precipitous slopes of
Mount Carrol or Mount Macdonald, as it is now called,
and the Hermit range. High above us to the northward the railway crept round the mountain buttresses,
protected for several miles by continuous shedding. We
must now ascend the spur of Mount Macdonald, where
it was completely forest-clad, and crossing its ridge try
to make our way into the Beaver Creek valley, on
the side nearest Mount Sir Donald. After zigzagging
about through dense tangled forest we came in about
one hour from the ford to the precipitous ascent.
The forest was bad enough where the slope was
gentle, but now, if there had been no forest whatever,
it would have been difficult to have made our way
upwards with a horse, but as it was, the whole
slope was covered with fallen trees in every stage of
decay, amid which large hemlocks, and cedars, and
spruce grew luxuriantly. There was an undergrowth
of rhododendrons, devil's club, and blueberry bushes,
and many of the big half-rotten logs were ready to
slide downwards with the slightest touch. Few people
in this country would believe that such a route was
possible for any beast of burden, but Ben was used
to this kind of work, and he and the horse seemed to
understand each other perfectly ; so between leaping
logs and making sudden rushes upwards, we gradually
found that we were rising higher and higher.

A series of spurts seemed to be the only way of getting the horse up, and when making one of these, the scuffle and smashing of branches was horrible. It seemed impossible that flesh and blood and skin could stand such work for long. Sometimes we could help matters by hewing a gap through a half burnt log, and making a passage for the horse. On one occasion I was ahead looking for the most open route, and was in the act of hewing a huge rotten log in twain when I heard the scuffle of the horse rushing upwards, then came a pause and a prolonged crash, and H. sang out, "The horse is gone!" "Gone! where?" "Down, of course," and down through the trees sure enough I could see the upturned tail and hind quarters, but the head was bent underneath. He was brought up in that position against a tree trunk, and remained perfectly motionless. I was very hot, and as Ben and H. were hurrying downwards I threw off my knapsack and sat down to see the result. When Ben got down to the poor beast he inspected his condition with much concern. I felt sure his neck was broken, and that he was stone-dead. Judge my surprise, when on Ben giving him a good shove, and holding on to the halter he tumbled over the log, alighted on his feet, just shook himself, and looked but little the worse for the tumble. Thanks to the diamond hitch the packs never came off, and as the horse rolled over and over they saved him from being torn badly by the snags and splinters through which he had fallen for about sixty feet. A little readjustment of the pack was necessary, and while they were at this, I swung myself once more into the straps of my knapsack, and returning to the big rotten log about four feet in diameter, had completed a passage through it before the horse had scrambled up so far.

Then he stood panting for a few seconds, and then came another rush amid crashing branches, and at last, at 10.30, the summit of the ridge was gained. Just at the top of the ridge the forest was more open and free from scrub, but the huge pines grew so thickly that we had to shape our course by compass.

After about a hundred yards of comparatively easy going we came to the descent towards Beaver Creek. Through the tree-tops we could get glimpses of the mountains beyond the valley, but any other outlook was blocked by the great pine trunks, as thick as the stalks in a corn-field and as big as ships' masts.

Bad as the ascent of the ridge had been, the descent towards the Beaver was far worse. The horse was scarcely able to move, without being in imminent peril of going head over heels all the rest of the way. After descending, with great difficulty, about two hundred feet, we halted and held a brief council of war. To take the horse further seemed impossible ; and no opening offered a view ahead. The best thing which suggested itself seemed to be to leave the horse on the summit of the ridge and ascend the *arête* of Mount Macdonald, till we could get a view over the valley which might suffice for mapping purposes. From the way the precipitous buttresses ran down from the main range, we knew that the gorges between would be invisible, and therefore we should give up some details we were anxious to obtain. To ascend the ridge offered partial success. Advance to the Beaver valley on this side, seemed well nigh hopeless. However, we were not going to abandon it without good reason, so Ben and I decided to go on and explore, leaving H. in charge of the horse. Throwing off our knapsacks and every other encumbrance, and taking nothing but the axes, which we found most useful in such a climb, we started downwards.

Seldom did our feet touch the solid earth. The whole descent for 1,000 feet was one series of gymnastic feats. Now and again a tree would have fallen in a right direction, and we could creep down its trunk for 50 feet or more.

The sound of the river grew louder and louder. At last the brink of a cliff was reached, and there, 100 feet below us, the Beaver swirled and foamed along.

It proved to be a swift torrent about 100 yards wide, with a considerable volume of water, making fording out of the question, and rafting seemed equally impossible. The land at the other side of the river was comparatively flat, and from it the forest-clad slopes curved up to the strange square summits of the Prairie hills. We had been told that some trail existed on this side of the river, and this hope led us on. Now we had proved to demonstration that no such trail was there. The place was absolutely impassable for a horse, so there was nothing for it, but to retrace our steps and scramble up through the terrible forest once more.

To retrace one's steps correctly in such a maze as this needs no little skill. Ben was an adept at such forest guiding; but when we considered we were within hearing distance we lifted our voices and got an answering shout from H., and thus found our way back to the horse. It was now too late in the day to think of any other move; there seemed to be nothing for it but to camp where the ridge was flattest. The worst of it was that we were a long way from a spring of water, which we had taken note of far below, and there was no grass or herbage of any sort to furnish feed for the horse. Having selected the best hollow we could find for the tent, Ben led the horse down to the spring for a drink, and then returned with the kettle full for our use. We picked the tops of rhodo-dendron bushes for the horse, but he would not touch them, so we had to spare him some scraps of biscuit; the poor beast seemed of such a friendly nature that he would go nowhere to look for food, but preferred

to stand as close to us as possible, looking into the
fire while we cooked.

The greatest danger of these forest camps is, of
course, that of setting fire to the dry lichens (which
hang like beards to the trees) and to the dry pine
needles with which the ground is thickly strewn, and
thus starting a huge conflagration. Another danger is
that throughout the living forest there are innumerable
dead trees, charred by former fires, which stand up
barely poised on end, and come crashing down at the
smallest touch, or after heavy rain, when the dead wood
becomes heavy from saturation, or when a puff of wind
with a thunderstorm sends them down like ninepins.
The great calmness of the Selkirk climate is proved
by the large number of these trees in a state of un-
stable equilibrium. On this occasion we had a good
illustration of what I have described. With our camp
things we brought a small net hammock; this we
slung between two big trees, and I lay in it to rest.
We had not examined the trees critically, but as I lay
and looked upwards I thought I saw the tree to which
my feet were slung, swing about in a suspicious
manner. I instantly jumped out of the hammock:
the tree was quite dead and ready to fall.

Both for fun and for safety we determined to have
it down; so getting the loop of a rope round the trunk,
we pushed it up as high as we could, with one of the
innumerable dead poles lying about. Then protecting

ourselves behind a big, healthy balsam we hauled away.
The tree gave two or three slight swings and then
came crashing down, making the earth tremble with
its fall.

A few little chipmunks and the tip-tap of the wood-
pecker were the only evidences of life. The heat in the
forest was very great, and the gentle evening breeze,
quite unfelt by us and imperceptible so far as any
visible movement of the tree tops was concerned, made
its presence known by a weird music, like the strains
of an Æolian harp, pervading all the air.

Though we were about 5,000 feet above the sea,
the forest was so sultry we had no need of our sleeping
bags. The trees get heated by the sun during the day
and so keep the temperature high during the night.
When we had safely picketed our steed close by the
tent, we sought the repose which he too needed badly,
after all his scrambles and tumbles.

CHAPTER XIV.

" The Pine,—Magnificent ! nay, sometimes almost terrible. Other trees, tufting crag or hill, yield to the forms and sway of ground, but the pine rises in serene resistance, self-contained."

RUSKIN.

Excursion up Mount Macdonald.—Recross Bear Creek.—Difficulties of the Tote road.

WHILE eating our breakfast at 5 A.M., we made plans for the day. Our great difficulty was the horse; we wished that he might vanish into thin air now that he had brought our provisions so far, for if we were rid of him, we could make a lengthened expedition from our present camp.

As, however, we did not see our way to help the poor beast into the spirit-world by slow starvation, and as he looked so patient and friendly, although he must have felt very hungry, we were obliged to give him due consideration. We decided, therefore, that he must not remain for another twenty-four hours without food, and that however high we might wish to climb during the day, we must return to camp in time to proceed down through the steep forest with the horse and regain the banks of Bear Creek before dark—that was the nearest place where anything of the nature of herbage could be found.

Packing the cameras, plane table, and some provisions in our knapsacks, we set off at six o'clock with the intention of getting up as high as was possible on Mount Macdonald, in the time we allowed ourselves. We followed the ridge upwards through the forest, which here was very dense and choked with logs and scrub, and, as the ascent was very steep, our progress was slow. Gradually the ridge narrowed, and as we skirted the brink of a great declivity overhanging Beaver Creek, we got through the trunks, a striking view of the river, meandering through its forest-clad valley, and of the eastern face of Mount Sir Donald with a hanging glacier close to the summit. Below this glacier the mountain looked nearly as difficult as on the side facing Glacier House, and the mountain spurs between us and it were high and precipitous. From these observations, the advice I would give to any one desirous of approaching the range from the eastward, is this —Raft the Beaver below its junction with Bear Creek: there is, I believe, a trail leading up the valley on the right bank of the river; follow this till opposite Sir Donald, and then re-cross the Beaver as best you can.

After about three hours' scrambling through forest, the trees became more gnarled and dwarfed in their growth, and the mossy hollows held banks of snow. Then we issued from the dark gloom of the pines on to a rocky ridge, into the blaze of the sunshine and the full view of a wide panorama including the most striking features of the Selkirks, the rocky peaks on the Watershed, the strange Prairie hills, the glaciers of the Hermit range, and beneath our feet the dark, deep ravine of Rogers pass, through which the railway wound its way, looking from our elevation like a thin hair-line twisting about with innumerable curves. The huge cliffs bounding the pass, and towering 5,000 feet above it, presented from this point of view a most imposing spectacle, while

> " On the torrent's brink beneath
> Behold the tall pines dwindled as to shrubs
> In dizziness of distance "

We ascended the rocky ridge, now gaining views down the cliffs to Rogers pass on the right, and then into the great ravine filled with a glacier on our left. At 10 A.M. we halted at a bank of snow, to partake of some food. The ascent was for the most part like going up a great staircase, the whole ridge being broken into angular blocks of rock. The rocks composing the ridge of Mount Macdonald are probably the most ancient of any we met with in the Selkirks. Mica schists predominated, these in many places were full of small garnets, in some cases replaced by curious pseudomorphs of mica. Professor Bonney, who kindly examined a specimen of the rock with many others which I brought home, expressed his opinion that it was no doubt of Archæan age.

At 11 A.M. we had reached the length of our tether, so far as time was concerned; so we halted on a little peak, which is plainly visible from the railway, 3,000 feet below, and setting up the plane table, I went to work at my mapping, while H. took a series of photographs and Ben prospected for specimens of rock. It was a most lovely day, and the view over the Beaver Creek valley to the Prairie hills, and away to the whole range of the Rockies, was most splendid. The further we advanced, however, towards the summit of Mount Macdonald, the more the buttresses flanking the main range closed in on each other, and offered less favourable positions for delineating their outlines.

We now became more impressed than ever, with the great difficulties which would have to be encountered, in any attempt to travel up the valley, at the foot of the spurs from the main range, and it seemed to us, that in order to get to the eastern base of Mount Sir Donald, the best route of all would be, to cross the watershed direct from Glacier House by one of the cols near Eagle peak. These we hoped we might have time to explore.

At 12.30 P.M. we began to retrace our steps, and after a delightful rock-scramble in the bright, clear air, we re-entered the forest, and plunged and slid and clambered down through the everlasting scrub and the tangle of dead and living trees.

It was easy enough to keep the right course when ascending, as all ribs of the mountain converged to the *arête*. Descending was quite another matter, and ere long we became convinced that the ridge we were following was not the one we had ascended by. Being completely enveloped in vegetation, we could see nothing that was a hundred yards distant in any direction. The sky was only visible in small patches. We first thought we had diverged too much to the right, and made a most difficult traverse to the left; then by my aneroid I knew that we were below the level of our camp, so we crossed back to the right, and ascending a couple of hundred feet, found our camp and our horse all safe and sound. We halted to boil the kettle and have dinner, and then making up the packs, we stowed them on the horse and began the descent to Bear Creek, sliding and smashing through dead branches just as before. Ben showed the most consummate skill in being able to keep a grip of the halter and stop the horse whenever, after jumping a fallen log, he seemed in danger of going all the rest of the way in a single tumble. The shades of evening were closing in when

the more level forest near the creek was reached,
and just before dark we were on the margin of the
river.   The packs were quickly off, and the poor
cayeuse lost no time in beginning his evening meal
off the rich tufts of grass, which he seemed to appre-
ciate after his fast of thirty-six hours.   The river
was now so much "up" with the melting of the
snow during a long, hot day, that fording was quite
out of the question.   The water was pouring over
the tree we had used as a bridge—we could see that
it was over four feet deeper than when we had crossed
before, and the volume of water passing down was therefore more than twice as great.   This was Saturday
night; we were anxious if possible to get back to
"Glacier" for Sunday, but there was no use thinking of fording till the chill of night had sealed up
the snowfields, and so reduced the waters to a lower
level.

Close to the river we pitched our tent.   The blueberries furnished us with a luxurious meal, and as
we had eaten all our bread we took this opportunity
to make a good-sized loaf with flour, soda, and
tartaric acid, and baking it in an oven of heated
stones, over which we built up a fire of fragrant
cedar logs.   When the loaf was safe out of the oven
we carefully extinguished every spark of fire, and,
turning in, slept soundly, though the torrent was
making the very ground tremble with its roar.

Sunday, August 19th.—We rose at 3.30, and not waiting to cook breakfast lest the river should begin to rise, we packed up our goods, and driving the horse into the water, carried the packs on our shoulders across by the tree. The river had fallen five feet during the night.

Arrived at the other side, we adjusted the packs on the pack-saddle, and with our knapsacks, containing cameras and instruments, on our shoulders, we struck up through the forest, and on reaching the tote road halted for breakfast.

As we faced up towards the great gorge of Rogers pass, the torrent, railway, and tote road were found to draw near together, so we determined to follow the tote road as long as possible. All along we had found it much cumbered by fallen trees, but after breakfast our troubles began in earnest—the further we got into the gorge the more the forest was devastated by fire. Here and there the charred trunks were smouldering; the air was choked with smoke; and at every twenty yards a huge tree lay right across the trail. Forcing our way through scrub, and often obliged by fallen trees to make wide detours, so as to get past one log which lay across our path, was most heart-breaking work; still we progressed slowly.

Then we came to a very wild stream, dashing right down in a series of cascades towards the torrent several hundred feet below. The bridge that once had existed was gone, and how to get past seemed an insoluble problem. Like all other mountain streams in these regions, every pool was choked up with fallen logs. Throwing off my knapsack, I scrambled, partly over logs, partly through water, to the other side. I saw that the next pool below offered a better chance for the horse, so shouting to Ben to this effect, I proceeded downwards. The roar of the cascade prevented my words being heard, and just as I reached the lower pool I saw Ben in the agonies of trying to steady the horse on the slippery logs; then quickly followed a splash and a scramble, and the next instant I saw the poor beast taking a somersault down the cascade to the pool at my feet. If he plunged at all, he would probably fall to a pool lower down, so seizing the halter I held on to him. He gasped and panted a good deal, but I did not let him get up till Ben had come down to my help, and then we landed him in safety: beyond a few scratches, no harm was done. Soon after this our plucky little steed made a splendid clean leap over another watercourse, and then we got on to the tote road once more.

Our progress was now slower than ever owing to fallen trees, fire, and smoke. We soon came to the conclusion that our only chance lay in striking up straight for the railway, which was about 300 feet above us. All this portion of the railway is covered by snow sheds, so we had to select a line of ascent which would

lead us to one of the few openings, and this involved
a scramble up through fallen timber and slopes of
stones lying at the highest angle possible for stability,
and ready to move downwards at the slightest disturbance.

After severe panting and scrambling, we at last
got to within twenty feet of the track. One rush
more, and we would be up! Ben mopped the perspira-
tion from his face, and led off for the last scramble,
but at this moment the horse put his foot on a slippery
plank and stumbled. I was beside Ben, and clutched
the halter to help him to hold the poor beast up,
but it was no use—his whole weight was on us. The
halter was jerked from our grip, and toppling backwards he went down twenty feet, alighted on the
top of his head, and then over and over in a cloud
of sticks, stones, and dust, for full one hundred feet.
It was too horrible—I felt certain he was killed,
but at a hundred feet he came to a sudden stop, and
we saw a slash-up of dust run along the slope!

A winter avalanche had carried away a lot of
telegraph wires, which were now lying amongst the
débris, and into these wires the horse tumbled, and
one of them coming across the packs, and (fortunately
for him) not against his body, brought him to a
stand.

On reaching the poor animal, we found him in a
kneeling position, with his nose buried in the dust,

and beneath him a confused mass of branches affording no foothold.

There was a terrible strain on the wire, and we feared at any moment it might break. Our first work was to cut the rope and remove the packs; then to hew away with our axes at the branches beneath him, while H. hung on to the halter, lest he should make a fresh start down the slope. When we found the horse could get his legs to the ground, we scraped a kind of track ahead of him, and in about twenty minutes everything was ready for a plunge upwards. With the spring of the wire (which was hitched against the back of the saddle) this was accomplished, and then, minus the packs, the poor beast gained the level of the railway in safety. H., Ben, and I carried the packs up on our shoulders; and then, after a short halt, we spliced the synch rope, got the packs once more into the diamond hitch, and entering the snow shed shaped our course for Rogers pass. For about three miles we enjoyed the cool shade of the sheds, and though we were alarmed at one time, by a tea-train from Vancouver rushing past and leaving us no more than two feet space clear of the horse, and on another occasion by a stray locomotive, which, by the way, vouchsafed us no warning whistle; we reached Rogers pass, and halted for lunch. As the West-bound Express would soon be coming along, we removed the packs from the horse. H. and Ben then led him on to "Glacier," four miles distant, and I followed in the train with the packs. So, after a hard day which we had commenced at 4 A.M., we were still not too late to hold our appointed evening service, enlivened by the singing of some well known hymns. Mr. Bell-Smith, who was not only a painter, but a musician, had spent some of his spare time during the week in forming a choir and practising hymns, which were now heartily sung by the inmates of the little inn, with no accompaniment save Nature's own music—the roar of the glacier torrent outside.

CHAPTER XV.

"The high hills are a refuge for the wild goats, and the rocks for the
conies."—PSALM cix. 18.

Start for Asulkan Pass.—Roast pork.—Mountain goats.—A narrow
bivouac.—Reach the Geikie and Dawson glaciers.—Thunder and
lightning.—A morning visitor.

As our days in the Selkirks were now numbered
it behoved us to waste no time. Accordingly, early
on Monday morning, we set to work and got our
packs ready for a fresh start. The photographs had
to be looked after and fresh plates put in. Our
purpose now was to cross the Asulkan glacier pass
and camp in the valley beyond. Thanks to the
excellent work done by four men during the past
week the trail was clear, and the glacier stream
bridged, so that we could take the horse about four
miles up the valley. For the rest of the journey
we should take the packs on our backs. This being
so, H. determined not to take his camera, but as I
was particularly anxious to secure good photographs

of the Dawson range and Geikie glacier, I packed my half-plate camera, which was much lighter than H.'s, and my "detective," into my knapsack with the surveying instruments and various items of clothing. When to these were added the cooking utensils, which fell to my lot, the pack weighed about forty pounds. Between blankets, waterproof sheet, and provisions, a rifle and ammunition, Ben and H. had about an equal share.

We packed as much of these things as possible on the horse, and took a tent to pitch at the furthest point to which he could go.

Charlie, whose business it was to watch the white stones in front of the inn, was off duty after the last train had come and gone, so he came with us to fetch back the horse, and our old friend "Jeff" came too. Immediately after the departure of the last train we fixed the packs on the poor cayeuse, who was much rested after a night in the stable and a good feed, and we started in Indian file up the Asulkan valley. The fallen trees had been sawn by cross-cuts, a path excavated from the hill side; an excellent bridge, perfectly safe for the horse, spanned the torrent, and was a most splendid illustration of rough-and-ready engineering skill. Our next crossing was made by the log by which we had got over before, and where the horse had to wade. The final crossing was near to where the stream issued in a wild rush from its cañon. As it was late in the evening the stream was very high, and the water flowed over the single tree trunk that had been felled for a bridge.

Here we had to unload the packs and carry them across on our shoulders, with difficulty steadying ourselves while the water went over our boots. "Jeff," in trying to follow, was swept away and, after disappearing under some logs and being nearly drowned, he considered himself fortunate to regain the safe side of the stream, and returned to his comfortable home. Charlie, too, objected to the tree trunk, so we left him to wait there while we took the horse on a little farther, our main object being to advance our camp as far as possible, but the trail now commenced to ascend the hill side in a kind of staircase, inaccessible for the horse. After some scrambling we retraced our steps and pitched our camp on a gravel flat in the river bed ; then sending the horse back across the stream, we said good-bye to him and to Charlie, as they set off on their way home.

It was now nearly dark, so as Ben said the cook had given him a piece of pork all ready to be eaten, we went at our supper without delay. The pork did not seem very nice, but we did not think much of that, at the time—it was so much food to work on, which was the main point. We had, however, ample opportunity to reflect on that supper, for we were

all nearly poisoned, and in a very ill condition for
carrying our heavy packs next day.

We were glad when morning came, and swinging
our heavy packs on our backs, we pushed up the steep
mountain side, noticing on the way a great number of
marmots.   Then after an hour over fatiguing moraine
we crossed the glacier, and halted near the last
vestige of vegetation to boil the kettle and have some
bread and tea.   After this we all felt a little more
cheerful, but our ascent to the col over the snow-
covered glacier was at a regular snail's pace.

It was just noon when we reached the summit
of the pass and found ourselves in a world of snow.
The thermometer, in the shade, registered 60° Fahr.,
and the glare of the sunshine was but little mitigated by
the smoke which rolled up from the burning forests.
Towards the Illecellewaet valley smoke rendered
photography impossible, but the air was clearer
towards the Dawson range.   Our first thought was
to look out for goats, so we descended the snow-
slope to the southward with caution, and watched the
grass slopes for some sign.   On the snow below us
were some objects which, at first sight, we thought to
be stones, but as we drew nearer we discovered that
they were eight fine goats, lying on the snow.   We
threw the packs off our shoulders at once, and crouching
amongst some stones discussed who was to have the
shooting.   We were all anxious for the fray, but as Ben

had carried the Martini and ammunition all the morning, I said he had best right to the sport, so off he started. H. and I sat on a crag watching for the result. Ben first ran quickly down the snow, and then getting amongst some rocks, crept cautiously to within about 200 yards or less of the game; nearer than that there was no cover whatever. Puff! went the little cloud of smoke, and we saw the ball plough up the snow about six inches over the back of one of the goats. We grew excited—puff! again the elevation was too great. Now the goat stood up, and the next ball knocked up the snow between his legs. We were getting frantic but dare not stir. It was evident that Ben, well used to his own Winchester, could not manage our rifle at all.

The goats were now all standing and seemed quite puzzled at the disturbance. We hoped the next shot might do for one; but no! The effect was a general stampede. Crack! crack! seven shots, and they were all out of sight round the buttress of the mountain. Ben attempted to follow but quickly returned to us rather crestfallen. I am proud to say we suc-ceeded in keeping our tempers, and fumed only inwardly. We were just thinking of resuming our journey when, high up on some rocks at the opposite side of the glen we were in, we saw a goat making his way upwards, and stopping every now and then to have a look around. H. said he would go after him;

so taking the rifle, he descended to the snow slope, and crossing it, gained the rocky cliffs opposite, and though he could not now see the goat we could, so we directed his course by pointing. The goat now got on to an overhanging cornice of snow, and it was most ludicrous to see H. peering about to try and get a sight of his quarry, and the goat craning out his neck to see where H. was, but keeping carefully out of sight himself.

Then H. from below the cornice, caught sight of the top of the goat's head and horns at a distance of about 200 yards. He fired a most hopeless shot which awakened a thousand echoes, and the goat turned round and strolled away to a higher point. After a desperate climb H. got on to the snow cornice, and we could see the goat a few hundred yards ahead, anxiously awaiting his approach. H. caught sight of him, up went the rifle, but the goat depressed its head at the same instant, and walked along the face of some inaccessible-looking rocks as though he had been on a well-made road. We saw that this kind of a hunt promised very little result and tried to signal to H. to come back. But no! on he went, so when I had completed a sketch, Ben and I shouldered our packs and descended to the grass slopes to look for a camping ground.

It was no easy matter to find one. We needed but a very small patch of level ground for a sleeping place, but

" Taking the rifle, he descended to the snow-slope."—P. 206.

nothing of the sort existed, and the further we descended the steeper it got. The mountain was so planed down by snow-slides in the spring, that everything like a knob, or hollow, or level was completely rounded off. We dared not set down a pack without securing it with an ice-axe ; even the smallest object showed a tendency to start off and tumble into the valley 2,000 feet below.

The Geikie glacier filling the bottom of the valley presented the most wonderful appearance. I never before saw a glacier so completely broken up into pinnacles of ice by longitudinal and transverse crevasses crossing each other. It presented the appearance of some basaltic formation with the blocks pulled a short distance asunder. A good deal of snow lay on its upper portion, and showed us what difficulties we must have encountered, had we continued our descent by it from the great Illecellewaet snow-field. The light was very good, and though a little smoke had drifted over the mountains into the valley, I succeeded in getting a very good half-plate negative of the glacier, with Mount Fox beyond. An hour had gone by since we had seen the last of H. and the goat, and as the climbing he had undertaken was as bad as any man need desire, we began to grow anxious. As we were also hungry we took advantage of a little dwarf scrub to collect materials for a fire, and with flour and Brand's Extract of Meat we concocted thick soup and enjoyed an invigorating meal. Two hours had now gone by and yet no sign of H.; dreading an accident we strolled upwards and soon were delighted to see him a long way up the mountain side, but descending towards us with his pack, to regain which he had had to make a long detour. He had got into the most break-neck precipices after the goat, but fired only two fruitless shots. Ben now took the rifle, and went off to seek better luck in the opposite direction, with only similar result.

Meanwhile H. and I sought for a camping ground, but finally had to set to work with our ice-axes and after an hour's digging, succeeded in cutting a notch in the hill side spacious enough to lie in. There was nothing for bedding save the dry earth, which, when softened by the axes, formed a nest by no means to be despised. We laid out the sleeping bags on the level we had made, and pegging the waterproof sheet to the upper edge of our little cutting, considered our camp absolutely perfect. It was now nearly dark, and when Ben turned up, we cooked a good supper and slipping under the sheet were soon in dreamland.

The descent to the Geikie glacier promised to be most precipitous, so we determined not to take our camp any further, but to descend, if possible; cross the glacier, and from the slopes of the Dawson range beyond take such observations and photographs as would enable me to plot out the south side of the

range, extending from Mount Bonney to the main
watershed of the Selkirks.

In the morning when we crept from beneath our
sheet the sun had not yet risen.   The eastern sky was
however rosy with the dawn; so also were the higher
snow summits, and in their brightness formed a striking
contrast to the cold grey of the glaciers as they wound
round the bases of the mountains into the dark gloom
of the valley.   The air vibrated with the music of the
cascade that splashed down the mountain side near our
bivouac, and the cry of marmot answering marmot in
the crags above warned us that a new day was breaking
and that it was time to be up and doing.

Breakfast over, we packed up our instruments, and
taking rope and ice-axes, prepared ourselves for the
descent.   At first there were no difficulties whatever;
then we came to crags from which everything in the
shape of a grip had been knocked off by avalanches in
the early spring.   We had to choose our direction with
the greatest care, and found the slopes of *débris* ever
ready for a fresh slide.   Finally we reached some alder
scrub, down which we slipped and slid, and then
stepped on to a heap of avalanche snow lying on the
glacier.   According to my aneroid the descent from
our camp was just 2,000 feet.   To get the exact width
of the glacier I now measured a base-line of 300 yards,
and setting up the plane table at either end took a series
of observations.   The glacier proved to be 1,000 yards

P

wide. While I was at this work H. and Ben went
prospecting along the cliffs, and to Ben's great joy
struck on a rich lode of galena in the mica schist.
We discussed what the mine should be called, but
never came to any conclusion, and the inaccessibility of
the valley will I fear make our find scarcely worth
naming. A row of fine specimens were laid out on the
snow for my inspection; the bright, shining cubes looked
most picturesque. Nevertheless I for one did not care to
add lead ballast to my cargo over the pass, and said so.
Ben however spoke up like a man, and said he would
carry it himself, and he did; but, as a natural consequence, H. and I had to carry an extra load of our
camp material and ere we got home Ben was nearly
dead beat with the weight of his new-found riches.

For the present we "cached" our specimens and
proceeded to cross the glacier. It was something
like the passage of the Mer de Glace from the
Montanvert, but near the further shore we had to
zigzag about for nearly an hour, and creep astride on
thin edges of ice, ere we could reach the moraine.
Where a spring burst from the mountain side we
halted for lunch. Though the heat was most oppressive, as the sun poured straight down into the valley,
there was an icy chill wind blowing down from the
snow fields. We were alternately roasted by the sun
when we sought shelter from the wind, or chilled to
the bone when exposed to it. A boulder on the

moraine was a point I had observed from the opposite side, so here again I set up the plane table, while
Ben and H. went prospecting. I suppose it was the
first time that a miner's pick clinked on the Dawson
range, but though there were a good many signs in
the mica schist that seemed hopeful, nothing of value
was found. We then ascended the moraine of the
Dawson glacier and examined the peaks surrounding
it. Mount Donkin might be easily scaled from this
side, so might Mount Fox, but the peaks at the head
of the glacier presented most precipitous faces, and we
knew from former observation that their gentler slopes
rose from the continuity of the great Illecellewaet *névé*,
and should be attacked from that direction.

Having sketched and photographed in all directions
it was now time to re-cross the glacier. This we did
with ease by a line below our former route, where the
crevasses were less open, and after a further examination of the galena lode, we commenced the steep climb
to our bivouac. For some distance we clambered up a
narrow gully where alder bushes gave good hand
grips, then we climbed the rocks, and reaching the
slopes covered with blueberry bushes, paused to collect
fruit for supper. Here Ben distinguished himself
notably and retrieved his honour as a sportsman.
Three fine grouse got up and flew in different directions; Ben followed them, and in three successive
shots with the rifle killed them. On reaching our

bivouac we were not long in consigning the three
grouse to the pot, and we finished up our supper with
squashed blueberries, sugar, and chocolate.

The sun had not yet set, and as we lay enjoying
the lovely view and the cool mountain air, so refreshing after the broiling heat of the valley and our
stiff climb of 2,000 feet, a distant rumble of thunder
struck on our ears. The sky was clear except for
thin wreaths of smoke drifting over the peaks to the
northward and giving to the sun a lurid glare. Now
we noticed a denser veil of cloud deadening the blue
and drifting rapidly over the sun, which became
blood-red. Denser and denser grew the drift. There
was no actual cloud; none with a sharp defined
margin as one is accustomed to see, but a dense veil,
shading from dull red to inky black. Suddenly it
blotted out the lower portions of the valley, and at
the same instant a flash of silvery white lightning[1] shot
from its midst, while the mountain walls reverberated
with crashing thunder.

The storm was approaching our position very rapidly.
The lightning flashed almost incessantly, and seemed to
strike the mountain sides just at our level. We felt
there was no retreat, and no shelter but our thin sheet
of oiled calico. Ben cheered us up as we shoved the

---

[1] The *whiteness* of the lightning always struck us as remarkable
This no doubt was due to the great dryness of the atmosphere in these
regions. This was also evidenced by the strength of the sun's actinic
rays, rendering no exposure too rapid for a successful photograph.

rifle under cover and pitched our axes away into the
short scrub, by tales of this sort : "It was just this
time two years that three men I knew, &c., &c.,
were caught in a storm like this and poor —— was
killed dead," and so on.

The storm had now reached Mount Donkin just opposite us; big drops began to patter on our sheet;
then came a blinding glare of lightning, and almost
simultaneously a bang like a cannon shot. That
was enough for us, we wished to see no more, so
huddling into our bivouac we folded up the sleeping
bags as tight as possible and lay on them; should they
get wet a miserable night was in store for us. Down
came the rain in a perfect deluge, but by lying on our
backs and keeping the sheet up with our knees we
caused most of it to run off. Our clothes got pretty
well soaked, but the sleeping bags kept quite dry.
Happily for us the worst of the storm clung along the
opposite side of the valley and was now passing away.
One howling blast of wind and all was over.

We now stepped out into the cool evening air; it was
dim twilight. The sun had set, and with much satisfaction we watched the great masses of clouds piling themselves away over the snow-field heading the glacier, and
a growling in distant valleys was the storm's last adieu.
Our fire had, of course, been extinguished, everything
was dripping wet, cascades innumerable were dashing
down their turbid waters. The wet made the air feel

still more chill, so we turned into our blankets and soon got warm and shortly afterwards fell asleep.

Thursday morning dawned clear, and ere sun-rise we were having a comfortable breakfast of tea and grouse. Though the sky was cloudless, there was a good deal of smoke in the air and the sun rose through it like

"A full-sized mountain goat inspecting us."

a glowing copper disc. We were sitting close round the fire, when Ben, who was facing me, made a sudden gesticulation and pointed to me to look round. I did so, and there, within five yards, stood a full-sized mountain goat inspecting us with the greatest interest. He was fairly puzzled to know

what kind of beings we were. The rifle was near
me under the sleeping bags, just where we had put
it when the thunderstorm was coming on. I instantly seized it, but the lock was so choked by
the dry earth that the lever refused to work. Franti-
cally I pulled away to try and get a cartridge in, but it
was no use. The goat, deeply engrossed in my move-
ments, took a step aside to see more distinctly what I
was at. Then Ben came to my assistance and between
us we got the breech open and the cartridge in. The
goat, by this, had strolled down into the gully close by,
and when we advanced to the edge, we saw him going
at his best speed up the rocks about 300 yards away.
I chanced a running shot and saw the ball knock up
the stones near his tail.

With the feeling that we had been insulted by that
goat we slung our packs on our backs. Ben, having
loaded himself with galena, could not take his share
of the camp things, so between cameras, instruments,
cooking utensils, &c., my pack weighed, as I found
by putting the things together on a subsequent
occasion, forty-two lbs. All the hard angular items
in it made it a very tiresome one to carry, especially
when descending the steep moraine where the foot
hold was loose.

We had to ascend about 1,000 feet to the col, and as
we went, we looked out for H.'s belt and knife that
he had thrown down on the spot where we waited for

him three days before, when he went off to hunt the
goat. We had some difficulty in finding the place, but
the sketch I had taken helped us, and by consulting it
we got the true line to the spot and found the knife.

Then we toiled up the snow slopes, and on reaching
the col, found that a bear had slept there on the snow
the evening before ; there was an exact cast of his
form and his footprints leading to and fro. I suppose
the heat of the valley, or possibly the shots he heard,
sent him upwards to this very airy situation for his
night's slumbers.

When we came near to the place where we saw
the marmots, H. shot one and I shot another, so we
were well off for supper, and then commenced the
descent to our tent on the river flat below. We had
to cross a cascade, and as it was now late in the day,
the stream was swollen to its utmost. The trail led
us to a place where we could not cross without going
into water above our boots, which Ben and I did, but
H., wishing to get across dry, clambered down to a
lower pool, where a log offered a possible chance of
getting over. H., with the utmost confidence, sprang
on to the log, which instantly revolving on its axis,
precipitated him, pack, marmot, rifle, and all, into the
cascade, and the worst of it was that he had on him,
slung by a strap, our mountain aneroid, and it *never
went again*. Fortunately he escaped being rolled by
the torrent over a fall of 100 feet ; so beyond the

wetting and the damage to the aneroid, and some
skin off his elbows and nose, he was nothing the
worse.

On reaching our tent we found the river so high
that had we wished to proceed to " Glacier" before
nightfall we could not have done so ; we therefore lit
the fire, and made a good supper of stewed marmot.

On the morning of the 24th, at **7** A.M., we left
our tent, packed up on the home side of the river,
and taking our knapsacks, walked down to " Glacier,"
which we reached at 9.   Later in the day Ben returned
with Charlie and the cayuse, and fetched down the
camp.

As we had now carried packs on our shoulders for
ten consecutive days, and had had no leisure to work
up observations, we determined to take two days
rest.   And there being nothing on hand in which Ben
could help us, we said good-bye, and he returned to
Illecellewaet.

CHAPTER XVI.

"The broad column which rolls on, and shows
More like the fountain of an infant sea
Torn from the womb of mountains by the throes
Of a new world."   BYRON.

Last ascent in Selkirks.—Golden City.—Up the Columbia.—Lake
Windermere.—Across the Rockies.

DEVELOPING photographs in the wine cellar and plotting out mountains, glaciers, and streams on our map,
was now the order of the day. Part of the 25th
was devoted to this work, and then we walked up
to make some observations on the glacier. On August
13th we had taken up a large auger, bored a series
of holes, and set up a row of poles across the glacier
to test its motion. We should have visited them
before now but that we had no spare day to do so.
On gaining the top of the moraine we found that
all the poles had fallen, owing to the surface-melting
of the ice under the powerful summer sun. By cutting
steps, we gained the surface of the ice, and found the

lower parts of the holes, and were able to set up a few of the rods again. The motion in twelve days seemed to be—No. 1 pole, near moraine, 7 feet; No. 2, further out, 10 feet: centre of glacier, 20 feet. The snout of the glacier showed evidence of retreat, for there were two rows of boulders in front of it. The outer one, about sixty feet from the ice, seemed to have been dropped the previous year; the inner row during the present year. As a test like that used in the valley of the Rhone glacier, I tarred some of the boulders in closest proximity to the ice. The retreat from these marks may be observed by future travellers.

When we were coming down the Asulkan valley on the previous day, we had noticed in some mud the footprints of a very large bear, which Ben pronounced to be a "silvertip." When he was going up to fetch the tent, he saw that the beast had traversed this same track again. It was therefore evident that it was the habit of the animal to come down to pick berries and return up the valley. Accordingly we determined to try for a last chance at a bear. There was a gentleman staying at Glacier who had come each year for twenty years to hunt in the American mountains and prairies, so he and I determined to go up the valley in the evening and lie in ambush for the beast; he had his favourite Winchester, and I took our Henri-Martini.

Having met with fresh tracks of the bear, we chose
a strategic position commanding a view of the path
for 100 yards, and we had the torrent between us
and it. If the bear was wounded and came for us,
we were thus sure of time for a couple more shots.
We dare not speak, and lay quite still. The sun
set ; the stars came out ; dark night made everything in the forest gloom invisible, no bear put in
an appearance, and as we were not prepared to lie
out in the cold till daybreak, we had to give up our
hunt, and return with an empty bag to "Glacier."
It is quite possible the creature winded us, and took
some other route home.

The 26th was Sunday, and as we determined that
we must say farewell to the Selkirks on the 29th,
it behoved to us to make one more ascent and gain
some point in the centre of the district we had sketched
out, whence we might tie all our bearings together
by a complete circle of angles. The mountain rising
immediately behind Glacier House offered the most
favourable conditions, so packing up the large plane
table and cameras, and accompanied by another
F.R.G.S., who, with his brother, turned up unex-
pectedly at "Glacier," we devoted the 27th to this
excursion. As Captain W. is a specialist in surveying, we were most fortunate in having his help, and
not only was the day spent usefully, but as the
distance was not great, we had ample time to en-

joy all the beautiful aspects of the panorama at our leisure.

Our course lay at first up through tangled forest, and besides the ordinary difficulties of clambering over fallen trees, an additional one presented itself in resisting the temptations offered by the blueberry bushes, which were covered with fruit in full perfection. Then we rose to grassy Alps where cow bells ought to have been ringing; then up rocks with banks of snow, and finally on to a rock *arête* where we set up the plane table at an elevation of 3,700 feet above the railway, and 7,804 feet above the sea. We were mid-way between the two big mountains of the district, Sir Donald and Mount Bonney, and by the level of the plane table we judged Mount Bonney to be the higher of the two; however in my map I have adhered to the observations of our aneroid taken on the latter mountain's summit. The view over the glacier world was as fine as could be desired, and for every reason I commend this climb to all travellers who stay off the train at Glacier, and desire some insight into the glories of the Selkirks. Below us to the northward, on a shoulder of the mountain, and embosomed in forest, was a perfect little gem of a lakelet. As we had heard that such a lake existed, Capt. W. had brought his fishing rod, so after we had sketched in the whole panorama, taken bearings and photographs, we descended to the grass slopes near a rill for lunch, and then made our way down steep rocks and through forest to the lake.

The glen above the lake which sloped down to a little beach, was nearly free from timber, and covered with rich grass. On all other sides the pines rose from the lake's edge in a dense wall, and were reflected in its still waters in minutest detail. Above them the snow-clad peaks of the Hermit range formed a serrated line against the sky. These, too, were reflected in the limpid water, and the stillness of the surface was so complete that all these reflections came out in a photograph. It seemed a pity to disturb its peace and stir up the mud, but the chance of a swim and the delights of a bath after the heat of the day could not be resisted, so in we went, and splashed about to our heart's content. After this we tried a few casts with the rod, but though the most enticing flies were put up no fish rose to them. This might have been on account of the disturbance we had made, but I incline to the belief that the reason why the fish did not rise was because there were none there. I could not even find any water snail or living creature in the shallow water. For more than half the year the lake must be a solid block of ice, so the absence of animal life is not to be wondered at. After a halt of about an hour we struck once more into the forest and left the lake calm and bright, and peaceful as we found it ; reflecting

the blue of heaven and the dark pine trees in its
depths.

According to the aneroids, three of which we had
with us, this little lakelet is 1,600 feet above the
railway, and down this 1,600 feet we now struck in
as straight a line as we could. The blueberries
again proved a cause of delay—it seemed impossible
to get past them, but at last (when we could eat
no more) we found ourselves on the railway, and a
mile along the track brought us to Glacier House.
The only thing that troubled us that night was,
that our last climb in the Selkirks was over, and all
that remained for us now to do, when we had our
last observations worked up was to pack our goods
and prepare to leave.

We had hoped very much to have had time for
another scramble on Mount Sir Donald, but there were
two other projects before us which would occupy
whatever spare time remained. One was to ascend the
Columbia river from Golden City to the lake from
which it flowed and so get a view of the whole flank of
the Selkirk range; also we had made an appointment
with Mr. MacArthur, the Government Surveyor in
the Rockies, to join him and see something of the
glacier peaks near the summit of Hector pass.

Nearly the whole of the 23th was spent in packing.
All our heavy luggage we despatched to Montreal,
and though Mr. MacArthur had promised to supply

the camp outfit when we joined him, we retained the small tent and sleeping bags in case we might need them.

The East-bound Express arrived in due time on the 29th, and at 3 P.M., bidding adieu to " Glacier," to our worthy host, and to " Jeff" and the bear, and the patient cayeuse, we took our seats on board, and started for Golden.

Soon after crossing Rogers pass we saw the last of Mount Sir Donald and the other peaks of the central range, and then descending the gradients towards Mountain Creek we were quickly enveloped in dense smoke arising from the burning forests.

> " Forests were set on fire—but hour by hour
> They fell and faded—and the crackling trunks
> Extinguished with a crash—and all was black."

For miles the magnificent timber was blazing away, and occasionally the wind drove hot puffs across our faces, like the blast of a furnace, as we stood on the platform of the car. The manner in which the fire worked was in some cases most peculiar. A huge cedar seemed to be all on fire inside, the flames shooting forth from openings all the way up the trunk, where branches had been broken off, like as from windows in a burning house. When the fire was near the railway the train went slowly, lest any injury should have happened to the track, but when clear of the sparks and smoke we shot on again, and leaving the ravine of the Beaver, were once more in the valley of the Columbia.

Amongst the passengers from Glacier was Mr. Irwin, the parson in charge of the whole district. He had just turned up after a ride of 1,200 miles through the mining camps, &c., in this district. He seemed to be the right man in the right place, and was thoroughly appreciated by all the section men and others along the line. At Donald we had time to visit his little wooden church. A carved pulpit and lectern had recently been presented by the workmen on the line. With their own hands the carving had been executed, and the engine-drivers, not to be outdone by the others, presented an organ. Golden City, our destination for the night, was in his district, and as, in consequence of his three months' absence in other parts of his parish, divine service had not been held there during all that time, he asked me to hold a service there that evening, and telegraphed on to get the congregation together.

When H. had crossed this country four years previously, Golden City had such a bad name that he and his companions made a detour of ten miles to avoid it. He was much interested in thinking over the change that time had wrought. A murder had been committed just before our arrival, and the people were as much incensed at the occurrence as they would have been in the most favoured community.

Q

On arriving at Golden City we put up at one of the two wooden inns. After supper the congregation began to assemble. There were a good many men, who sang the hymns most lustily and I for the first time experienced the encouragement of a "Hear, hear," during a very short sermon. Next morning one of my congregation joined me on the river bank, and asked me was I a High Churchman or a Low? My reply was, "That all depends on what you are, for if you are one of the two you would most likely call me the other." He thought for some time, and then said he was not sure which he was, and did not know what was the difference between them, but he had always heard that his father was a High Churchman. As the possibility of a definite answer was now moved a generation further back, we changed the subject and discussed the future of Golden City—the great probability that its inhabitants would experience something like what happened to the people whom Noah did not take with him in the ark.

The heat during the day in the Columbia valley was intense, but an intelligent inhabitant told me that though their winter climate is much mitigated by the warm Chinook winds, still, in consequence of sharp frosts occurring at night even in July and August, vegetables cannot be grown with any certainty of a crop.

The great chance in favour of Golden being a success is that it is situated at the lower termination of the navigable portion of the upper Columbia. Below Golden, the Columbia, as it sweeps round its great northern bend, forms a series of rapids, and a great many lives have been lost in attempting to go down from Golden by boat to the Arrow lakes. Part of the journey—*i.e.*, as far up as Canoe river, which leads to the Athabasca pass—was regularly traversed by the *employés* of the great fur companies in the days gone by, and for this river journey, skilled *royageurs* were needed; that from Canoe river to Golden is a much more dangerous portion. One man who had seen the very first wooden house set up at Golden City, told me that numbers of miners, prospectors, and settlers used to arrive there from the eastward, and considering the Selkirk range impassable, would build rough boats or rafts and start down the Columbia. No one ever knew what became of them, and the belief was that most of them perished. There are now two little stern-wheel steamers on the upper Columbia, the *Duchess* and *Marion*, and as the *Duchess* was now making her last trip for the season, we were fortunate in securing berths on board for a voyage of about 100 miles to Lake Windermere. The *Duchess* draws about eighteen inches of water, but as the river was falling very low there were fears that she might get aground. The *Marion*, which would take her place, draws only six inches, but has no cabin accommodation. On August 30th

at noon we drove across the shingle flats formed by
the shifting of the Wapta river, and reaching the
Columbia went on board the *Duchess*.   The captain,
the "purser," three hands, and a China boy for cook,
formed the crew.   The berths were as comfortable as
one could wish, and the cooking first-class.   The other
passengers were an English party going up to visit
their relations, who were settled in the Kootenay
country, where Mr. Baillie-Groman was constructing a
canal connecting the head-waters of the Columbia and
Kootenay.

The heat in the valley was very great, and the
mosquitoes a perfect plague, so we were glad when
our paddle began to revolve, and the breeze caused
by the motion gave us some relief.   The river scenery
was very pretty rather than grand; clumps of pines
varied by an abundance of birch, aspens, and other
deciduous trees, were all reflected in the still water.

The current being gentle we made good progress,
and following the course of the river more into the
centre of the valley, we obtained fine views of the
Selkirk range on our right and the bare cliffs of
the Rockies rising on our left.   The contrast between
the two ranges was very great.   The Selkirks rose in
swelling domes of dense forest, range after range of
heights, the inner ones alone being rugged and snow-
capped.   The Rockies, on the contrary, rose abruptly
from terraces of white silt, showing the various levels

occupied by the river in ancient days, and but sparingly dotted with trees ; to great walls of pink and white limestone and rugged snow-seamed crags.

From the Selkirks, many large tributaries, of which the Spillamachene river was the most important, came to swell the waters of the Columbia. From the Rockies, for over eighty miles, we saw nothing of the sort. Dry torrent-beds with a mere trickle of water met the Columbia. From the boulders and gravel heaps near them, we could judge that they had their spring season of heavy floods but no permanent source of supply. Where the valley widened out there were extensive lagoons, or "slews," separated from the river by narrow alluvial banks, pierced by occasional openings. These lagoons were fringed by great meadows of sedge, reeds, and other water-plants, and the smooth surface, occasionally ruffled by the rising of a flock of geese or duck, reflected the pine forests and the mountains to their summits. Besides geese and ducks, large groups of cranes were occasionally met with on the sand spits, large kingfishers darted along the margin of the stream, and ospreys, whose nests were formed of huge bundles of sticks on the top of pine-trunks, the upper portion of which had been carried away by storm or lightning, were nearly always in sight. On one tree-trunk a fine white-headed eagle stood inspecting us, but quickly moved off when a bullet from the captain's Winchester whizzed past him.

All day long we steamed up stream, and admired the reflections and the mountains which, though much obscured by smoke, formed the great feature of the view. A great deal of the smoke, the Captain told us, was owing to a fire set going by prospectors. Some men discovered a mine a few weeks previously in one of the Selkirk glens and came down to Golden to lodge their claim. Two other parties hearing of their find, determined to "jump the claim," and set off up the river. One set came up in the *Duchess* on her last trip, others went by land. The first party on reaching the valley set the forest on fire behind them, so as to keep the others out, hence the volumes of smoke now filling the air and obscuring the whole panorama. When night closed in we had to tie up to the trees on the river-bank, and we resumed our voyage at daybreak. The river got shallower as we advanced, and some of its curves proved most difficult to turn: one of these we only got round by the help of a rope and our capstan. At another place we touched at, there was an encampment of Shushwap Indians; they had come down the river to pick and dry berries for their winter use. They had ponies with them, and one little cedar-bark canoe of the most picturesque model, bow and stern being closed in like the toe of a sharp-pointed slipper. The berries were sewn up into packs covered with birch bark, and were all ready to ship on board the steamer. While these were being loaded we strolled on shore. One girl had a little papoose slung on her back in a wooden case, its little black eyes sparkling brightly inside a cage of mosquito netting, which covered its face. The cartridge cases round the men's shoulders were gaily adorned with pieces of brass and coloured cloth, and altogether the group was extremely picturesque.

Further up we came on an encampment of Kootenay Indians, but they were nearly quite naked, and looked more wretched than the Shushwaps.

A few years since, the Indians made these excursions for berries to the same grounds, when they little knew what steamboats were like; now they had so accommodated themselves to the advance of civilization that instead of driving horses laden with packs of berries, to their homes, they placed the packs in the steamer, and when they reached their destination other Indian women produced their invoices and took charge of the goods.

The last tributary coming from the Selkirks, Toby's creek, was a turbid mountain stream. Beyond this the Columbia was crystal clear and very shallow. The gravel was pushed into heaps by the salmon that used to spawn here in myriads; but we were informed that they never come there now, and over-fishing at the

mouth of the great river is the reason assigned as the
cause of their disappearance. Whether salmon come
there or not, I never before saw such multitudes of
trout.   They shot out from beneath the steamer in
dense shoals, and varied in size from about half a pound
to three pounds.   We touched the bottom and almost
stuck fast several times, but owing to the captain's
great skill in selecting the deepest water by zigzaging
about amidst the gravel heaps, we got through all right
and entered upon the broad expanse of Lake Windermere.
On a headland jutting into the lake were the remains of
an old fort of the Hudson's Bay Company, and all
along its shores, benches and terraces of gravel and
white silt, dotted with fine groups of Douglas firs,
illustrated the high levels once occupied by its waters.
At 6 P.M. we reached the head of the lake, the
present termination of navigation, and tied up to a
little wharf called "Sam's landing" until morning.
After a delightful bath in the clear water we strolled
up to some bluffs and sketched the landscape.   The
wide levels of the lake stretching away for miles;
the reeds growing out of the water and giving
shelter to thousands of wild ducks and geese whose
quackings and cacklings filled the air; the white
benches of silt bounding the lake in all directions,
and the lofty peaks of the Selkirks, showing as a
serrated line of purple against a crimson sky, combined
to make a most impressive scene.   If our steamer

"A little wharf called 'Sam's Landing.'"—P. 232.

had possessed any small boat we should have felt inclined for a duck hunt; but the captain told us he had tried it several times with little or no result, for the ducks would not rise, and swam ahead, faster than the boat could get through the reeds, and always kept out of sight. We had a large party on board the *Duchess* that evening, for the friends who came to meet those who had come up the river with us stayed on board, their ranch being too far distant to reach that evening. At daybreak they left us, some in a waggon, some riding and shortly afterwards we let go from the wharf and started on our return voyage. We had now seen the east flank of the Selkirk range, and its termination in the Kootenay plains, and gained some idea of the big valleys through which rivers flowed with the main drainage of the glaciers. From the Kootenay lake to the southward, the Selkirks have been penetrated by prospectors and hunters for many miles. Here Indian packers and hunters may be found, and game is abundant.

The descent of the Columbia was naturally much more expeditious than the ascent. We bumped along on leaving the lake over the gravel of the salmon-beds, but thanks chiefly to the velocity of the stream, did not stick fast. We had plenty of time for sketching and writing up notes, the want of active exercise seemed the only tiresome feature of this trip; so we were always glad when the steamer stopped to take on board fire-wood, for in heaving the logs into her from the bank we obtained some refreshing exercise. We took two days to go up, one day sufficed for coming down the river, and as we approached Golden City we became a little anxious about catching the train, for we hoped to cross the Rockies that same evening. Fortunately, on reaching the station, we found that the train was behind time. After waiting for an hour the Atlantic Express arrived, and taking our seats on board we were in a few minutes rattling up through the cañon of the Wapta. The roar of the torrent in its rock-bound channel sounded louder in the darkness, and now and then the glare of burning pine trees made its white foam seem tinted with deep blood-red.

When we had passed this way in July, though there was evidence of former fires, none were actually burning: now the whole air was charged with smoke, and the forest on the mountain sides was in many places blazing like a furnace. About midnight we crossed Hector pass, and shortly after 1 A.M., on September 2nd, reached Banff, where we got off the train, which went away on its course to the prairies, and we drove to an inn called the Sanatorium to get half a night's rest.

CHAPTER XVII.

> " Farewell to the mountains high covered with snow !
> Farewell to the straths and green valleys below !
> Farewell to the forests and wild-hanging woods !
> Farewell to the torrents and loud-pouring floods ! "
>
> BURNS.

Silver City.—Astray in the forest.—Lake Louise.—Mount Lefroy.—
Swamped in the Bow river.—Once more on the prairie.

FROM an artistic point of view it might have been
better that this chapter should have come nearer to the
beginning of my story than at the end, but as I have
adhered to the true sequence of events it must come
here or nowhere. If I told not of our misfortunes
I should leave out a little glimpse of the beauties of
the Rockies, which it was our good fortune to enjoy,
though other feelings somewhat spoilt the enjoyment
at the time.

Banff, where we now spent Sunday, August 2nd, rose
quickly into some importance once the railway was constructed, for there are picturesque stalactite grottoes with
hot medicinal springs in the vicinity. In one of these we

enjoyed a delightful bath, and I feel sure that if we had had anything the matter with us we should have experienced instant relief.

In consequence of the existence of these springs, and of the mountain scenery which surrounds this valley, the Canadian Pacific Railway Company built here a large and very picturesque hotel, managed in the best possible style, commanding the very best view, and provided with a doctor who tells visitors how to take the baths, and when they are sufficiently convalescent, to pay their bills and go somewhere else. Besides the Canadian Pacific Railway hotel, the "Sanatorium" is another excellent establishment. Then there are shops selling articles of vertu, posting establishments, several smaller inns, two churches, a *theatre*, boats to hire on the river, provision shops, post office, altogether it is a most respectable little town; and so interesting is the surrounding country that the Government have allotted it to the people for ever as a national park. Our chief reason for coming here was to meet Mr. MacArthur, to whom I had written ten days before.

We had no little difficulty in finding him, and when we did, he apologised for not having replied to my letter, but the fact was, that owing to the great prevalence of forest fires, he had not been able to make such progress with his survey as to be free to come with us to Mount Lefroy. We felt not a little disappointed, but there was no help for it, nor any use in wishing we were back again in the Selkirks, where the few days at our disposal might have been spent most profitably. My work being in the Selkirks specially, I had taken no trouble to make up the topography of the Rockies, except so far as was possible from the small scale-map of the Geological Survey. On that map Mount Lefroy was marked as the highest peak of the Rockies, 11,658 feet, but that map could not be trusted for details of routes, which we had not considered, as we expected to have been guided to the base of the mountain.

Now that matters had taken another aspect we deliberated for a few minutes, and then, as we were anxious to make the acquaintance of the finest mountain scenery in the Rockies, even if the ascent of Mount Lefroy was impossible, we decided to start off and make out our own way to the mountain. On this being settled, Mr. MacArthur was able to give us some help; he sketched out the mountain in my note-book, and explained that if we wished to ascend the peak our best chance would be by ascending the Vermilion pass. If, on the other hand, we felt satisfied with a near view of its most precipitous side we might approach it from Laggan, about forty miles from Banff, on the way to Hector pass.

He also told us that at Silver City a man named Joe Smith resided, who owned three horses, and that as he was usually on for any adventure, we ought to find him

out in the first instance. Silver City was eight or
nine miles from Banff, at the mouth of the valley
leading to the Vermilion pass and near Castle
Mountain station, on the Canadian Pacific Railway.
In our great innocence we asked if there were a
telegraph to Silver City so that we might communicate with Joe Smith. "Telegraph! what an idea!!"
so there was nothing for it but to make up our
packs and set off for Silver City in person.

Fortunately we had brought a tent and blankets,
now we had to purchase a kettle and frying-pan, some
bacon and hard sea-biscuits, and making them up in
two handy packs, we left our other luggage in charge
of the station-agent, and made inquiries concerning
the next train. We were told that a freight train was
due on its way to the West at 10 A.M.

Then a telegram came to say that it could not
arrive till 12, so we loafed about the town for two
hours. At 12 a telegram arrived saying that it could
not be at Banff before 3 P.M., and finally at 4 P.M.
it appeared in sight, and we took our seats in the
caboose with the conductors, and started for Castle
Mountain.

Half an hour's run brought us to the station, which
consisted of nothing but a water-tank for the engine,
and here we left the train. The mountains had at
this place receded from the river, leaving a level plain
surrounded by very bare castellated peaks, and on the

centre of this little plain we could see the rows of brown log houses forming Silver City.

Throwing down our packs near the railway we walked off to explore the "City." There was a marvellous stillness about it; no sign of any living thing!

We reached the grass-grown street, and looking into the empty houses heard no sound but the flapping of torn paper, which once bedizened the walls. One house we explored had evidently been a drinking saloon; there was some moth-eaten finery hanging about and an underground cellar. At last we espied a dog; it seemed inclined to be friendly, so hopes of Joe Smith rose in our breasts. Then we caught sight of a woman, and at the farthest end of the town found one house, in which were two young women, the sole inhabitants of Silver City! One was cooking, the other washing, and after a little conversation we found that they were the wives of men working in the section gang on the railway. They also told us that six men resided here, that they would soon arrive for their supper, and that we might stay and join them at it if we wished. We asked for Joe Smith, but to our chagrin heard that he had started six weeks before with his horses, on a prospecting tour to the southward, and was not expected back for another month. The dog we had seen was his dog, but it preferred the flesh-pots of the section gang to the hard life its master led in the mountains. For our night's lodgings they told us we might occupy any house in the "City" we liked. What a charming place some might think Silver City to be! House and land and "no rint!" Having no landlord to badger and boycott would be one drawback, and besides the loss of this diversion, there was an abundance of spiders; so notwithstanding the freedom of everything, we preferred pitching our own tent on the plains, outside the group of log cabins, and though there was nothing to make our bed soft, we scooped little hollows for our hips and shoulders and were quite comfortable at night.

The men at work on the railway returned about sunset, and after supper two of them strolled over to our camp fire and sat for an hour's chat. They were both Canadians, born of Canadian parents, and considered the railway work a very easy life. They were most intelligent, had plenty to say, and altogether were admirable specimens of humanity. We discussed the past history of "Silver City," which was brief, as may be supposed, and chiefly depended on a swindle carried out by some sharp mining agents. They knew where Mount Lefroy was, and could have pointed out its peak to us but for the dense clouds of smoke, which made everything indistinct. A party of Stony Indians with squaws, on a hunting expedition had passed up the valley just before us, and had gone away by the trail for the Vermilion pass.

If we had been earlier we might possibly have gone on with them. None of the railway men knew anything of these trails, and their advice was, not to try the Vermilion without some guide. Joe Smith had guided a party over the Vermilion into the Kootenay country in the spring, before the snow was gone.

As no guide was to be had, and as we did not care to run the risk of starvation by seeking a way up an altogether unknown valley, we decided that we must rest content with what we could see of Mount Lefroy by going up the line to Laggan, and from there approach its precipitous face *viâ* Lake Louise. Camped on the bare plain, 5,000 feet above the sea, and with no sheltering forest we found the night air very chill, but at 5 A.M. we got up, and taking our packs to the water tank on the railway we waited for the Pacific Express which being "on time" arrived about 6 A.M.

A run of thirty miles brought us to Laggan, which is a place of more importance than Castle Mountain, for besides the inevitable water-tank, there are sheds in which the two monster locomotives, kept for working the steep gradients to the westward of the pass, find a home. Here also reside three troopers of the red-coated police, whose business it is to watch the pass, and prevent liquor being smuggled into the North-West territories from British Columbia. The station agent, who by the way is a first-rate entomologist, a contractor for cutting trees in the forest for the railway, two women

R

and a child, with the section gang, made up the whole
population, and to the boarding house where all these
fed we made our way for breakfast.

A most excellent breakfast we got, and while eating
it we discussed our future plans. The contractor said
that while prospecting for timber he had been up as
far as Lake Louise, and advised us to follow the railway track for about two miles up the valley, and cross
the Bow river by the railway bridge, and then strike
up the mountain side through the forest. One of the
railway men said he had a skow on the river, and
though it was not very seaworthy, and the river was
very swift, if we chose to take the risk of the
passage we were welcome to it, and that by thus
crossing the river we would save an hour's walking.
We accepted his offer, and after breakfast took up our
packs and went with him to the river, which is here
a series of rapids, with one pool where the current
being less swift offered a possibility for a ferry. The
skow was just like a magnified pig-trough made of
rough boards and square at both ends. At present
it was sunk, and quite water soaked. We were not
long, however, baling it out with the help of our kettle
and frying-pan, and as the rapids commenced immediately below, we hauled it up along the bank as far as
possible so as to make a good start. Putting our packs
on the one seat in the middle, H. took one paddle
and I the other, and getting a good shove off from

the owner, by paddling with might and main we reached the other shore in safety, but drifted down about 100 yards while making the passage. Here, making fast the skow to a tree, we shouldered our packs and struck up the mountain side for Lake Louise. We could see the valley occupied by the lake quite plainly from the railway, but every one had warned us that when we entered the dense forest it was quite possible for us to lose our way. The last word from the station agent was, " Be sure whatever happens you keep to the left, for then you will meet the creek from the lake."

At first our course lay up through the blackened poles of a young burnt-out forest. Then crossing some swamps we entered the living forest, and though much more open and composed of smaller trees than what we had been accustomed to in the Selkirks, we had to do a good deal of scrambling over fallen logs. Soon we were shut out from all view of the outside world and had to consider our direction with more care. The ground being uneven it was not always easy to judge our direction by the lie of the land. But keeping the station agent's last words before my mind, whenever alternative routes opened ahead I always took the one to the left. H. now considered that I was going too much to the left and preferred bearing away to the right, and as there was no third person to decide who had the best of the argument, and as the compass could

R 2

not be appealed to owing to our not having mutually agreed upon bearings before entering the forest, he bore away to the right and I kept to the left. As long as possible we kept up communication by shouting to each other, and as we could not be far from the lake I felt certain that we must soon meet on its shore. In less than half an hour from hearing H.'s last shout I struck upon the stream coming from the lake, and in a few moments more stood upon its swampy shore.

I was quite unprepared for the full beauty of the scene. Nothing of the kind could possibly surpass it. I was somewhat reminded of the Oeschinen See in Switzerland, but Lake Louise is about twice as long; the forests surrounding it are far richer, and the grouping of the mountains is simply perfection. At the head of the lake the great precipice of Mount Lefroy stood up in noble grandeur, a glacier sweeping round its foot came right down to the head of the lake. Half way up the cliffs another glacier occupied a shelf, and from its margin, where the ice showed a thickness of about 300 feet, great avalanches were constantly falling to the glacier below. Above the upper glacier the peak rose in horizontal strata, the edges of which were outlined with thin wreaths of snow, to a gently sloping blunt peak crowned with a cap of ice. The mountains closing in on either side and falling precipitously to the lake formed a suitable frame to this magnificent picture.

"At the head of the lake, the great precipice of Mount Lefroy."—P. 244.

The lake was of the deepest green-blue, like those in
Switzerland, and the pine forest growing actually into
the water, clad the mountain sides in dense masses
wherever trees could find enough earth for their roots.
All this was reflected in the lake, which was barely
ruffled by little puffs of wind, now striking in one place
now in another, and causing the water momentarily to
sparkle in the sunshine.

There was a little too much smoke in the atmosphere
to make photography a success, but I took a couple of
views at once, for dark clouds heaving up from behind
the peaks looked ominous, and I feared a break in the
weather.

As shouting met with no response save the echo
from the rocky walls, I lit a fire and piled on a
quantity of green boughs, so as to make a good smoke.
If H. was anywhere in the vicinity I felt sure he
would turn up soon, so set to work hauling a few
floating logs together, and when three were lashed
side by side they made a famous raft, on which I
thought we might paddle our camp to the head of
the lake. Having paddled about a little to test its
stability, I began to feel hungry, so opening the pack,
I cut a few rashers of bacon, but had no bread or
biscuit to eat with it, H. having all the biscuits
in his pack; my pack consisted of the tent, one
sleeping-bag, the bacon, kettle, frying-pan, and
tea. His was made up of one sleeping-bag, a warm

rug, the thin oiled sheet, biscuits, and a tin of beef.

Dinner over, and very heavy, threatening clouds pouring down from the mountain heights and filling the valley, I thought it better to pitch the tent, and had to break up my raft, for the rope stretched between two trees was needed for fixing it. I had this scarcely done when there came a blinding flash of lightning and a crash of thunder, which seemed almost as if all the mountains were tumbling about my ears. Then came a fierce storm of wind down the lake. The placid surface was lashed into spindrift in an instant, and the wild crashing of trees as they fell before the blast, rendered the scene almost terrifying. Then flash succeeded flash in quick succession, but the wind storm only lasted about five minutes.

All this time I was crouching in the tent, which I had pitched just inside the margin of the forest, and did my best to keep the sleeping-bag dry, but what with fluttering and banging about, a good deal of wet got in. In about half an hour the whole storm was over, the lake assumed a dull, leaden hue, the clouds hung low, and the wild cry of a loon swimming far out from the shore, was the only sound that broke the stillness. The vegetation was all dripping, and as night approached it felt very cold. Tea without milk, and bacon without bread was not luxurious, but I had

"New views of strange castellated crags opened as I progressed."—P. 247.

to be satisfied, and wondering what had become of H.,
I turned into my sleeping-bag, but was too cold to
sleep much. Avalanches, too, were constantly falling
from the hanging glacier at the head of the lake.
Between 2 and 5 A.M., the period of the greatest cold,
they fell most frequently, at intervals of about twenty
minutes, making the whole valley resound as with
thunder. I was glad when day broke, and getting up,
lit the fire and made breakfast—tea and bacon once
more—then frying a number of pieces of bacon, I made
them up in leaves, and with this provision for the day,
started to walk round the lake and so reach the glacier.

Before leaving my camp I wrote a note for H. should
he turn up, and nailed it with one of the spare boot
nails I carried in my pocket to a tree close by. The
clouds still hung low, and it began to rain. There was
a kind of trail which I followed when possible, sometimes scrambling up into the woods and again wading
in the water. The mountain-tops were cut off by
clouds, but new views of strange castellated crags opened
as I progressed. It was 6 A.M. when I started. In two
hours I had reached a point not far from the glacier
stream, above which the moraine rose in high piles of
crags partly covered with vegetation. For more than an
hour it had rained steadily, and between it and the
reeking vegetation my clothes were as wet as if I had
fallen into the lake. I now saw the utter hopelessness
of further advance, the heavy rain clouds were coming

lower and lower, so making a few hasty sketches I sat at the foot of a pine tree, and eating some bacon, attempted a last photograph, and then turned my back on Mount Lefroy whose cliffs were now quite invisible.

On regaining the tent I made no delay, but packing it up with the sleeping bag inside, got the pack on my back. Owing to the wet it was about twice the weight it was before, and so disconsolately enough I started down through the forest to the Bow river. Not feeling certain that I should find the skow, I bore away to the left, and in due time reaching the railway bridge I crossed it, and arrived at Laggan about noon.

The police seeing that I was like a drowned rat took pity on me and lent me a change of clothes, so I turned out for dinner in riding breeches with broad yellow stripes down the legs, slippers, and an oilskin coat. I was a little anxious concerning H., and would have been more so only that the troopers had found evidence of his having returned to the skow, which was still fastened where we left it. There was a piece of white paper visible, fastened to a paddle, so H. must have returned so far, and then gone back up the mountain to look for me. I had left a note for him on the shore of the lake so it most probably was only a question of waiting.

About 3 P.M. a man came bringing information that H. was coming down to the skow, and all hands turned out in great excitement to see him attempt the passage single-handed. I came along too, as fast as the riding breeches would permit and brought the camera, so as to get a good photograph. One of the troopers ran off for his lasso, to catch H. if possible in mid-stream, and evidently to carry out the proverb concerning the " man that is born to be hanged," &c.

On H. reaching the skow the whole population of Laggan, including two women and one child, stood on the other side of the Bow. Carefully he placed his pack on board, then he began to haul the skow up stream and away from the rapids as far as possible; this was difficult and very tedious. All considered the odds decidedly against him. I shouted out to hold on till I could walk round by the railway bridge. But the roar of the waters drowned my voice. Then came the critical moment; he stepped on board, quickly coiled down the rope and shoved off—one frantic stroke of the paddle and the craft instead of going ahead, spun round like a top. It was now evident that all was up. The current made no delay, so I fixed the camera for one particular rock where the first wave of the rapids curled over. I felt sure that here the catastrophe would take place, but though I got a very fair instantaneous photograph, considering the weather it was a moment too soon. Passing over the first wave in safety the skow swept past the rock, but was swamped three seconds later. She did not roll over, but coming broadside on to a rock, filled and remained firmly jambed, with the river pouring over blankets and everything else.

The lasso was flung, but fell short. Then H. took to the water, and scrambling from boulder to boulder, the spray occasionally going over his head, he reached the shore. As soon as he regained breath, taking the end of some ropes we had tied together, he went into the rapids again, and making the end fast to the skow, he lifted the wet pack on to his back, out of which the water poured in a perfect stream. To carry the pack to land involved a real hard struggle with the seething water, but the beholders were much impressed, and one of the troopers remarked, "I guess that chap has got some sand in him," which I took to be no end of a compliment to H., and felt that whatever the "sand" might mean, it was well deserved. All hands hanging on to the rope, we hauled the skow to land, her symmetry being sadly marred by the process. H. now had to borrow a rig out, and we spent the rest of the evening in the police-hut playing whist and enjoying the good company of our kind hosts, while our clothes steamed away on a nearly red-hot stove.

H. had spent rather a dreary night of it too. On parting from me he bore away through the forest, and kept on mounting upwards till 4 P.M., when the thunderstorm came on. He got a glimpse of the lake when over a thousand feet above it, and it being easier to return to the Bow than to reach it, he struck back to the river and camped near the skow for the night. In the morning he started up for Lake Louise by the right route, passed me in the forest, and found my note.

That day, September 5th, was our last in the mountains. About midnight the Atlantic Express arrived, on board which we took our seats in the first-class car. As we had to pick up some things at Banff and again at Calgary during the night, we did not go into the "sleeper;" and when the morning dawned and our rather weary eyes opened, we were far away on the prairie. The Rockies and the Selkirks were but a memory of the past.

THE END.

RICHARD CLAY AND SONS, LIMITED, LONDON AND BUNGAY.

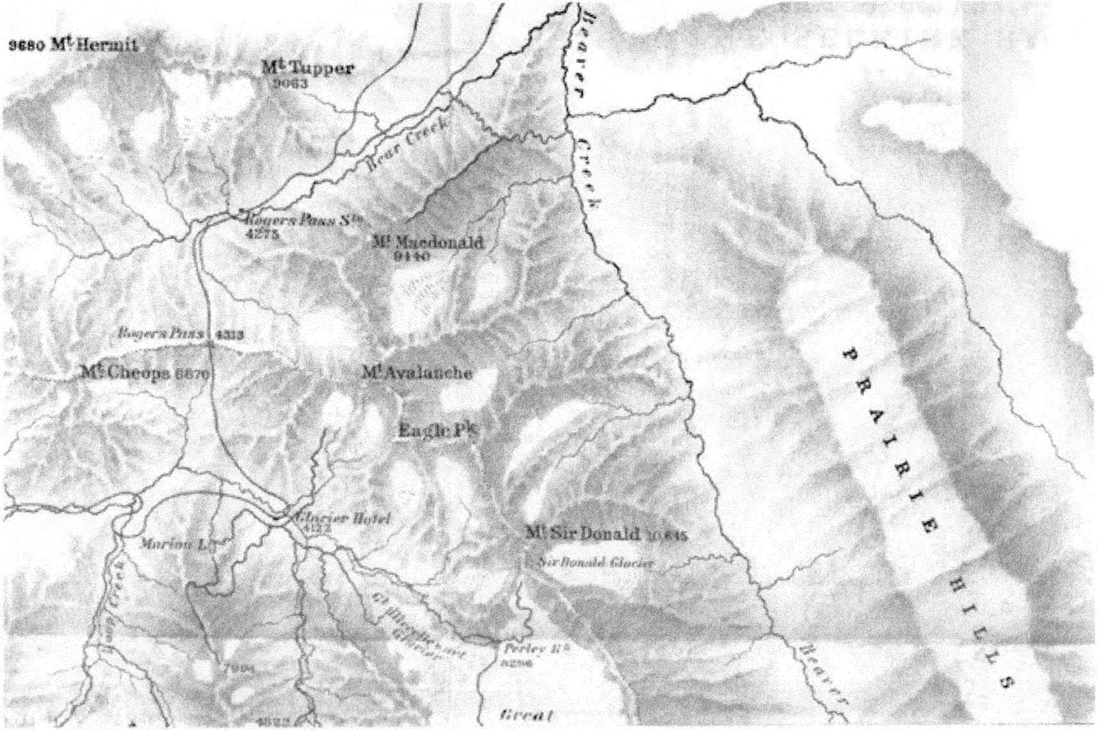
9680 Mt Hermit
Mt Tupper
9083
Bear Creek
Beaver Creek
Rogers Pass Sta
4275
Mt Macdonald
9440
Rogers Pass 4313
Mt Cheops 8870
Mt Avalanche
Eagle Pk
Glacier Hotel
4122
Marion L.
Mt Sir Donald 10,645
Sir Donald Glacier
Perley Rk
Great
Beaver
PRAIRIE HILLS

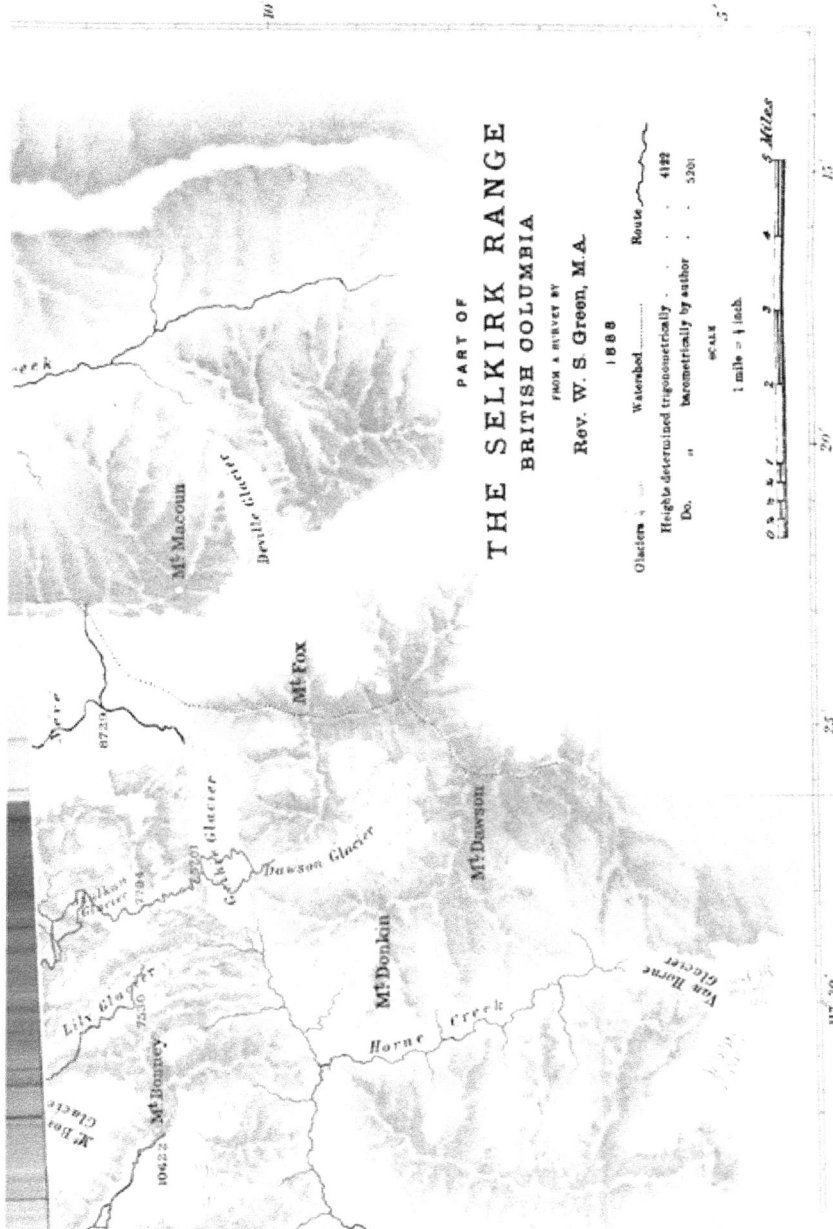

PART OF

THE SELKIRK RANGE
BRITISH COLUMBIA

FROM A SURVEY BY

Rev. W. S. Green, M.A.

1888

Glaciers
Watershed
Route

Height determined trigonometrically . . . 4192
Do. „ barometrically by author . . . 5201

SCALE

1 mile = ⅛ inch

0 ½ ¼ 1   1   2   3   4   5 Miles

Mt Macoun

Deville Glacier

Mt Fox

Mt Dawson

Mt Donkin

Donkin Glacier

Dawson Glacier

Lily Glacier 7530

Mt Bonney 10422

Mt Bos Glacier

Van Horne Glacier

Horne Creek

8720

7794

8500

THE SOUTHERN PART

OF

BRITISH COLUMBIA

Scale
English Miles

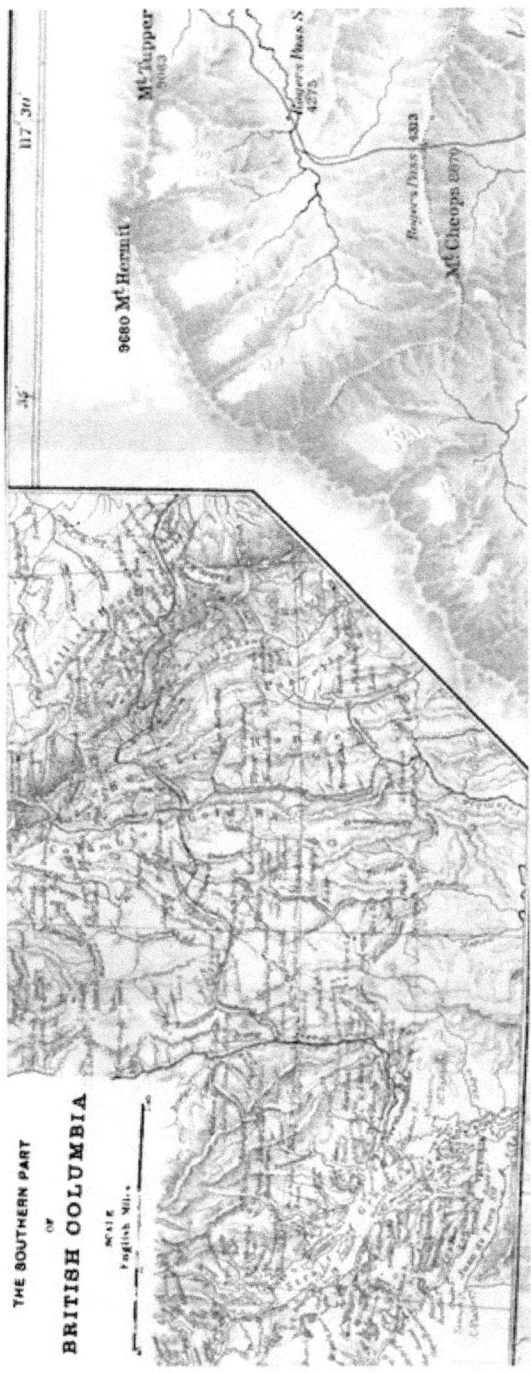

117° 30'

9680 Mt Hermit

Mt Tupper
9363

Rogers Pass St.
4275

Rogers Pass 4313

Mt Cheops 8970

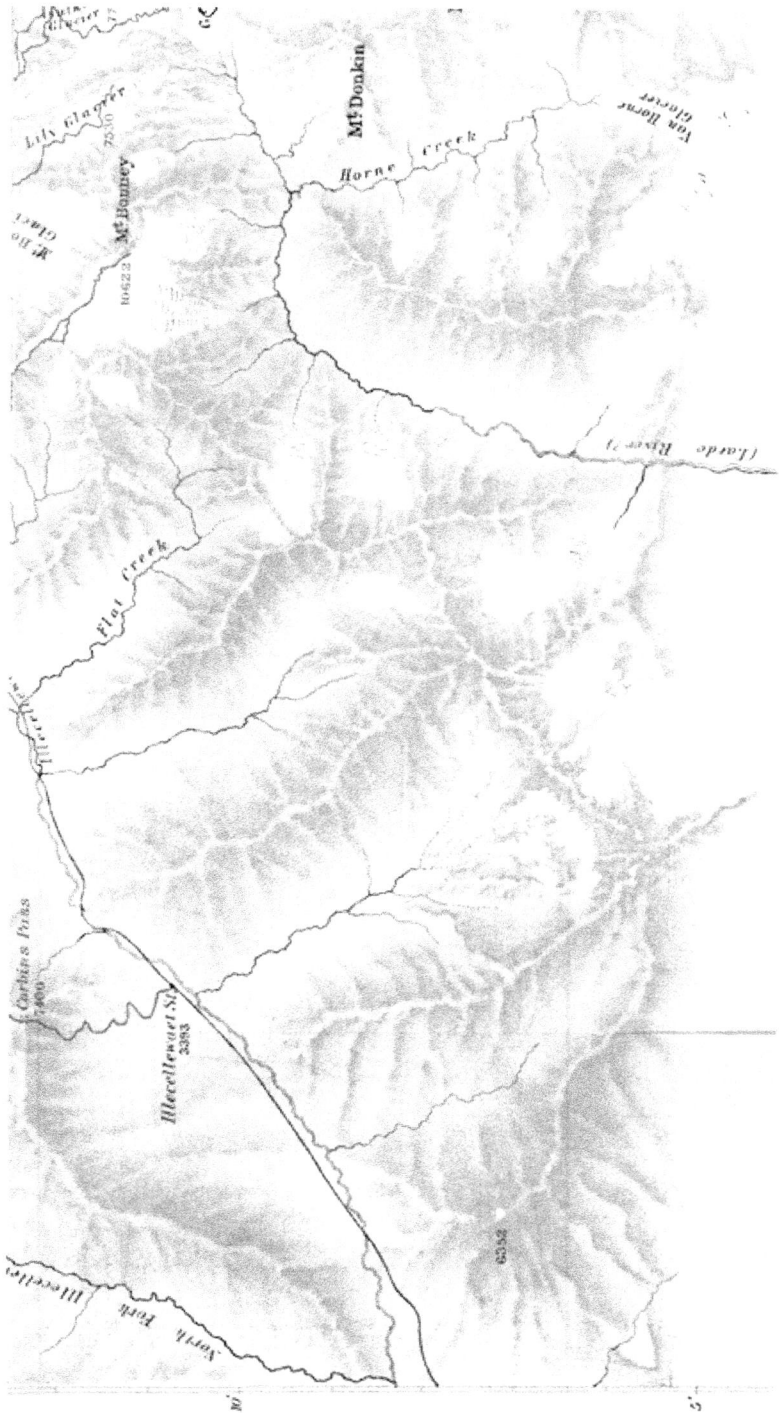

MICROSCOPICAL PHYSIOGRAPHY OF THE ROCK-MAKING MINERALS: AN AID TO THE MICROSCOPICAL STUDY OF ROCKS. By H. ROSENBUSCH. Translated and Abridged for Use in Schools and Colleges. By JOSEPH P. IDDINGS. Illustrated by 121 Woodcuts and 26 Photomicrographs. 8vo. 24s.

A TREATISE ON ORE DEPOSITS. By J. ARTHUR PHILLIPS, F.R.S., V.P.G.S., F.C.S., M.Inst.C.E., Ancien Elève de l'École des Mines, Paris; Author of "A Manual of Metallurgy," "The Mining and Metallurgy of Gold and Silver," &c. With numerous Illustrations. 8vo. 25s.

A POPULAR TREATISE ON THE WINDS. Comprising the General Motions of the Atmosphere, Monsoons, Cyclones, Tornadoes, Waterspouts, Hailstorms, &c. By WILLIAM FERREL, M.A., Ph.D., late Professor and Assistant in the Signal Service; Member of the American National Academy of Sciences, and of other Home and Foreign Scientific Societies. Demy 8vo. 18s.

PHYSICS OF THE EARTH'S CRUST. By the Rev. OSMOND FISHER, M.A., F.G.S., Rector of Harlton, Hon. Fellow of King's College, London, and late Fellow and Tutor of Jesus College, Cambridge. Second Edition, altered and enlarged. 8vo. 12s.

EARLY MAN IN BRITAIN AND HIS PLACE IN THE TERTIARY PERIOD. By W. BOYD DAWKINS, F.R.S., Professor of Geology in Owens College, Manchester; Hon. Fellow, Jesus College, Oxford. With numerous Illustrations. Medium 8vo. 25s.

ACADIAN GEOLOGY, THE GEOLOGICAL STRUCTURE, ORGANIC REMAINS, AND MINERAL RESOURCES OF NOVA SCOTIA, NEW BRUNSWICK, AND PRINCE EDWARD ISLAND. By Sir J. W. DAWSON, LL.D., F.R.S., F.G.S. With Geological Maps and Illustrations. Third Edition, with Supplement. 8vo. 21s.

THE THEORY OF THE GLACIERS OF SAVOY. By M. LE CHANOINE RENDU. Translated by A. WILLS, Q.C. To which are added the Original Memoir, and Supplementary Articles by Professors P. G. TAIT and J. RUSKIN. Edited, with Introductory Remarks, by GEORGE FORBES, B.A. 8vo. 7s. 6d.

GEOLOGY AND ZOOLOGY OF ABYSSINIA. By W. T. BLANFORD. With Coloured Illustrations and Geological Map. 8vo. 21s.

JOURNAL OF A TOUR IN MOROCCO AND THE GREAT ATLAS. By Sir JOSEPH D. HOOKER, K.C.S.I., C.B., F.R.S., &c.; and JOHN BALL, F.R.S. With Illustrations and Map, and an Appendix by G. MAW, F.L.S. 8vo. 21s.

Messrs. MACMILLAN & CO.'S PUBLICATIONS.

THE LIBRARY REFERENCE ATLAS OF THE WORLD. A
Complete Series of 84 Modern Maps. By John Bartholomew,
F.R.G.S. With Geographical Index to 100,000 places. Half-
morocco, gilt edges, folio, £2 12s. 6d. net.

*** This work has been designed with the object of supplying the
public with a thoroughly complete and accurate Atlas of Modern
Geography, in a convenient reference form, and at a moderate price.

PALÆOLITHIC MAN IN NORTH-WEST MIDDLESEX : The
Evidence of his Existence and the Physical Conditions under
which he lived in Ealing and its Neighbourhood, illustrated by
the Condition and Culture presented by certain Existing Savages.
By Jno. Allen Brown, F.G.S., F.R.G.S. With Frontispiece
and Eight Plates. Demy 8vo. 7s. 6d.

REPORT ON THE EAST ANGLIAN EARTHQUAKE OF 22nd
APRIL, 1884. Being Vol. I. of the Essex Field Club Special
Memoirs. By Raphael Meldola, F.C.S., F.I.C., F.R.A.S.,
&c., and William White, F.E.S. With Maps and other Illustrations. Cheaper Issue. Demy 8vo. 3s. 6d.

GILBERT WHITE'S NATURAL HISTORY AND ANTIQUITIES
OF SELBORNE. New Edition. Edited, with Notes and
Memoir, by Frank Buckland, a Chapter on Antiquities by
Lord Selborne, and the Garden Kalendar. With Illustrations.
Cheap Edition. Crown 8vo. 6s.

AN UNKNOWN COUNTRY (A TOUR IN IRELAND). By the
Author of "John Halifax, Gentleman." With numerous Illustrations by Frederic Noel Paton. Royal 8vo. 7s. 6d.

THE FERTILISATION OF FLOWERS. By Professor Hermann
Müller. Translated and Edited by D'Arcy W. Thompson,
Jun., B.A., Scholar of Trinity College, Cambridge. With a
Preface by Charles Darwin, F.R.S. With Illustrations.
Medium 8vo. 21s.

PROFESSOR CLERK MAXWELL, A LIFE OF. With Selections
from his Correspondence and Occasional Writings. By Lewis
Campbell, M.A., LL.D., Professor of Greek in the University of
St. Andrews, and William Garnett, M.A., Principal of Durham
College of Science, Newcastle-on-Tyne. New Edition, abridged
and revised. Crown 8vo. 7s. 6d.

LECTURES AND ESSAYS. By W. K. Clifford, F.R.S., late
Fellow and Assistant Tutor of Trinity College, Cambridge ; Professor of Applied Mathematics and Mechanics at University
College, London. Edited by Leslie Stephen and F. Pollock,
with an Introduction by F. Pollock. With Two Portraits. 2
vols. 8vo. 25s. Also an abridged Popular Edition. Crown 8vo.
8s. 6d.

SEEING AND THINKING. By W. K. Clifford, F.R.S. With
Diagrams. Crown 8vo. 3s. 6d. [Nature Series.

MACMILLAN & CO., LONDON.

www.ingramcontent.com/pod-product-compliance
Lightning Source LLC
Chambersburg PA
CBHW020506270326
41926CB00008B/764